Emergence of Chaotic Dynamics from Singularities

Pablo G. Barrientos
Santiago Ibáñez
Alexandre A. Rodrigues
J. Ángel Rodríguez

Emergence of Chaotic Dynamics from Singularities

INSTITUTO DE MATEMÁTICA PURA E APLICADA

Emergence of Chaotic Dynamics from Singularities
Primeira publicação, julho de 2019
Copyright © 2019 Pablo G. Barrientos, Santiago Ibáñez, Alexandre A. Rodrigues e J.
Ángel Rodríguez.
Publicado no Brasil / Published in Brazil.

ISBN   978-85-244-0490-0
MSC (2010) Primary: 37C29, Secondary: 37G10, 37D45, 37G20, 37C20, 37C23

Coordenação Geral   Nancy Garcia

Produção   Books in Bytes   **Capa**   Sergio Vaz

Realização da Editora do IMPA
   IMPA   Telefones: (21) 2529-5005
   Estrada Dona Castorina, 110   2529-5276
   Jardim Botânico   www.impa.br
   22460-320 Rio de Janeiro RJ   editora@impa.br

```
B275e      Barrientos, Pablo G.
                Emergence of Chaotic Dynamics from Singularities / Pablo G.
           Barrientos, Santiago Ibáñez, Alexandre A. Rodrigues e J. Ángel
           Rodríguez. - 1.ed. -- Rio de Janeiro: IMPA, junho de 2019.
                ISBN 978-85-244-0490-0
                1. Chaotic dynamics 2. Singularity. I. Título II. Série
           CDD: 515  CDU: 514.2
```

Contents

Poincaré (1890) was the first who pointed out the complicated dynamics that the existence of a homoclinic point could involve. Forty years later, Birkhoff (1927) proved that there were infinitely many periodic points, with arbitrarily large periods, in a neighborhood of a transversal homoclinic point q. Smale (1967) explained Birkhoff's results placing his geometrical device in a neighborhood of q, the so-called *horseshoe map*.

The horseshoe map is a diffeomorphism f defined on the sphere $\mathbb{S}^2$ with a compact invariant set Λ given as the product of two Cantor sets. In addition, the set Λ is hyperbolic, transitive and the dynamics of f restricted to Λ is topologically conjugate to a Bernoulli shift. This allows to prove that the set of periodic points of f is dense in Λ and also that Λ contains an infinite number of periodic points with arbitrarily large period. Both, diffeomorphisms like the horseshoe map and also those introduced earlier by Anosov (1967) on the torus $\mathbb{T}^2$, belong to a wider class of systems, the so-called axiom A, uniformly hyperbolic diffeomorphisms or, simply, hyperbolic diffeomorphisms. The non-wandering set of a uniformly hyperbolic diffeomorphism f decomposes as a pairwise disjoint finite union of basic sets, which can be seen, in the most tangled scenario, as the invariant set of a horseshoe map. The invariant set Λ of a horseshoe map contains an expansive dense orbit which implies exponential dependence on the initial conditions and, consequently, uncertainty. Mainly due to this fact, chaotic dynamics was understood as the presence of horseshoes. However, Λ is not an attractor and therefore it is not observable. We say attractor to refer to any compact invariant set whose stable set has either non-empty interior (or positive Lebesgue measure). A repeller is an attractor for the backwards dynamics.

Nowadays, at least for families of dissipative systems, *chaotic dynamics* is mostly understood as the *persistence of strange attractors*. An attractor is said strange if it contains an expansive and dense orbit. Expansivity means that the orbit has at least one positive Lyapunov exponent. Given a parametric family of dynamical systems, a property, as for instance the existence of strange attractors, is said persistent if it occurs for parameter values in a positive Lebesgue measure set. Persistence of a dynamics is physically relevant because it means that the phenomenon is observable with positive probability. As we will explain later, for families of diffeomorphisms on surfaces, homoclinic bifurcations are, as long as they involve the creation or destruction of horseshoes, a well-understood gateway to the emergence of persistent strange attractors. This fact allows to find, via Poincaré maps, vector fields in $\mathbb{R}^3$ with suspended persistent strange attractors close to a recurrent orbit. Hence, for families of differential equations in $\mathbb{R}^3$, a typical context for many real-world models, the persistence of strange attractors can be proved near many homoclinic or heteroclinic cycles. None of these configurations is easy to find in a given vector field and therefore, the presence of strange attractors is usually concluded by means of numerical methods. Singularities of a vector field, that is, the equilibrium points, are the only elements of the phase portrait that can be easily determined, sometimes even algebraically. For this reason, *the main goal of this book is to provide criteria for the existence of homoclinic cycles, and consequently the persistence of strange attractors, in generic unfoldings of some (non-hyperbolic) singularity of a vector field.* In short, we are proposing a feasible and manageable route to access the study of dynamic complexity in a given family of vector fields: the study of singularities of low codimension.

Singularities for which all eigenvalues of the linearization have a non-zero real part are said hyperbolic. Let us consider a family X_μ of vector fields on $\mathbb{R}^n$, with $\mu \in \mathbb{R}^k$, such that $p_0 \in \mathbb{R}^n$ is a hyperbolic singularity when $\mu = \mu_0$. Implicit Function Theorem and Hartman–Grobman Theorem imply that, for μ close enough to μ_0, there exists an equilibrium point p_μ, with $p_{\mu_0} = p_0$, such that the phase-portraits of X_μ and X_{μ_0} are locally equivalent. That is, given a family X_μ of vector fields, hyperbolic singularities, if exists, they are (generically) present in codimension zero subsets of the parameter space. In this sense we say that hyperbolic singularities are of codimension zero.

Dealing with non-hyperbolic singularities, the number of degeneracy conditions imposed, either at the level of linear or at the level of higher order terms, determines the codimension of the singularity. Let X_μ be a family of vector fields, with $\mu \in \mathbb{R}^k$, and assume that there is a singularity at p_0, when $\mu = \mu_0$, of codimension $c \leqslant k$. Codimension c means that, generically, the singularity persists for parameter values in a set of codimension c in the parameter space. Note also that singularities of codimension c are avoidable in k-parameter families with $k < c$, but generic if $k \geqslant c$. In short, the codimension of a singularity coincides with the smallest number of parameters which are required to give generic unfoldings. It is clear that the smaller the codimension the more abundant the singularity is.

Given a dynamical behavior, either chaotic or not, it is of great interest to determine the singularity of lowest codimension which unfolds that behavior. We can understand this singularity as the *seed* from which that dynamics emerges. Given a family of vector fields, we will want to study if it contains the good seeds to guarantee that a certain dynamics is present. From the perspective of applications, looking for seeds is a much more manageable task than trying to directly detect a certain global phenomenon. Bearing in mind this idea, seeds were found in Drubi, Ibáñez, and Rodríguez (2007) to prove that the coupling of simple dynamics can lead to the persistence of strange attractors: A thought-provoking idea in biology and in other branches of science. From a more theoretical point of view, linking singularities and dynamics allows to transfer the study of the dynamical complexity to the analytical hierarchy that the notion of codimension introduces in the set of the singularities.

In this book, we propose a journey through singularities, unfolding, bifurcations and strange attractors. The first stage is a walk through essential results in the context of diffeomorphisms. After that, we see how these results can be applied to flows of vector fields via Poincaré return maps. In this way, the persistence of strange attractors in families of diffeomorphisms which unfold a homoclinic tangency lead to the persistence of suspended strange attractors in families of vector fields. We explain how these attractors are formed from homoclinic bifurcations, mostly around a Shilnikov homoclinic orbit or around a bifocal homoclinic orbit, and from heteroclinic bifurcations as, for instance, in neighborhoods of Bykov cycles. All these global configurations will be introduced later. These structures are difficult to detect in a phase space, but we prove that they arise in generic unfoldings of certain singularities. As already mentioned, proceeding in this way, we provide results that allow to conclude the persistence of strange attractors from the presence of certain singularities in a given family.

The first example of strange attractor was the solenoid: a hyperbolic strange attractor built by Smale (1967) on the solid torus. From this example, the term strange attractor was used by Ruelle and Takens (1971) to provide an explanation for the nature of the turbulence. However, solenoids do not appear naturally in families of dynamical systems. On the contrary, other examples such as the attractors found numerically by Lorenz (1963) and Hénon (1976) arise in simple families of quadratic vector fields and diffeomorphisms, respectively. Because of their inherent dynamics, these attractors looked like strange attractors, but they did not have the properties of hyperbolic attractors. This is why during the

70s, some relevant questions were raised: Do non-hyperbolic strange attractors really exist?,...If they exist, How abundant are they?, How do they arise?,...Or better, Which bifurcations are related with their emergence?

Obviously, the answer to the above questions had to be sought outside the set of uniformly hyperbolic systems. It was learned very soon that these systems were not dense in the set of regular dynamical systems. In the space of C^1 diffeomorphisms defined on a compact manifold M with dimension $n \geqslant 3$, open sets of non-uniformly hyperbolic C^1 diffeomorphisms were constructed from the existence of robust *heterodimensional cycles* (see Abraham and Smale (1970) and Simon (1972)). A diffeomorphism has a heterodimensional cycle associated with two transitive hyperbolic sets if these sets have different indices (dimension of the stable bundle) and their invariant manifolds meet cyclically. These cycles are not possible for $n = 2$, but open sets of non hyperbolic C^r diffeomorphisms, with $r \geqslant 2$, are obtained from robust homoclinic tangencies between the invariant manifolds of a non-trivial basic set (a horseshoe), Newhouse (1970). Namely, in this paper, Newhouse proved that there exists an open set $\mathcal{U}$ of C^r diffeomorphisms with $r \geqslant 2$ where diffeomorphisms having homoclinic tangencies associated with hyperbolic periodic points are dense. Hence, the uniformly hyperbolic diffeomorphisms are not dense. Each of these open sets $\mathcal{U}$ is called a *Newhouse domain* and it contains a residual set $\mathcal{R}$ such that every $f \in \mathcal{R}$ has infinitely many periodic attractors or repellers (*Newhouse phenomenon*) Newhouse (1974). In fact, the open set $\mathcal{U}$ is arbitrarily close to each C^r diffeomorphism that has a homoclinic tangency Newhouse (1979). New and maybe more clear proofs of Newhouse results were presented in the book of Palis and Takens (1993) where also can be found a parametric version of this result, see Robinson (1983).

All in all, homoclinic tangencies are a route of access to new non-hyperbolic dynamics. Therefore, they could well be the gateway to the existence of strange attractors. On the other hand, a phenomenon such as the Newhouse could occur for infinite strange attractors. In this context, Palis formulated several conjectures which have been collected in Palis (2000), a part of them were aimed at answering the previous questions.

The first analytical proof of the existence and persistence of non-hyperbolic strange attractors was given in 1991 for the Hénon family

$$H_{a,b}(x, y) = (1 - ax^2 + y, bx), \tag{1}$$

in a remarkable work authored by Benedicks and Carleson (1991). Consider generic C^∞ one-parameter families f_μ of surface diffeomorphisms that unfold homoclinic tangencies between the invariant manifolds of a saddle-type periodic point for $\mu = \mu_0$. Using the ideas in Benedicks and Carleson (1991), Mora and Viana (1993) proved that strange attractors or repellers are persistent in f_μ for

μ close to μ_0. The starting point of the proof in Mora and Viana (1993) is a renormalization which allows to show that, as it happens with (1), the family f_μ is a good unfolding of the quadratic limit family $f_a(x) = 1 - ax^2$, in a neighborhood of the tangency point. Attraction or repulsion depends on whether f is dissipative, or not, at point p. As Palis had conjectured, homoclinic bifurcations (cycles and tangencies) become a generic mechanism to create non-hyperbolic dynamics and, in particular, for families of dissipative diffeomorphisms persistent strange attractors. Together, genericity and persistence imply the abundance of strange attractors in dynamical models. Some advance on this conjecture has been done in Crovisier and Pujals (2015) and Pujals and Sambarino (2000).

Regarding coexistence, an analogue to the Newhouse phenomenon was proved in Colli (1998) for strange attractors. Colli proved that in the Newhouse domains $\mathcal{U}$ of C^r diffeomorphisms for $r \geqslant 3$ large enough, diffeomorphisms exhibiting infinitely many coexisting strange attractors form a dense subset of $\mathcal{U}$. In fact, this result is achieved from the existence of residual sets of one-parameter families unfolding homoclinic tangencies analogous to the parametric version of the Newhouse phenomena. Namely, any generic one-parameter family $(f_\mu)_\mu$ unfolding a homoclinic tangency at $\mu = \mu_0$ has an open set I of parameter value arbitrarily close to $\mu = \mu_0$ such that f_μ exhibits infinitely many coexisting strange persistent attractors for μ in a dense set of I. Although each of these attractors is persistent in the family $(f_\mu)_\mu$, like in the Newhouse phenomenon, the coexistence of infinitely many of them is not persistent. At this point an open question remains: Are there k-parameter families with infinitely many attractors, either strange or not, for values of the parameter in a positive Lebesgue measure set $E \subset \mathbb{R}^k$? In a new conjecture, Palis claimed that the Lebesgue measure of the possible set E is generically zero for families of one-dimensional dynamics and surface diffeomorphisms. Although the Palis conjecture remains still open, some advances in the opposite direction have been made by Berger (2016, 2017) for families of surface endomorphisms (in fact, local diffeomorphisms) and higher dimensional diffeomorphisms.

The above mentioned strange attractors are one-dimensional, namely, they are the closure of the one-dimensional unstable manifold involved in a homoclinic tangency. To obtain two-dimensional strange attractors, that is, strange attractors where the sum of the Lyapunov exponents is positive, diffeomorphisms defined on a three-dimensional manifold and with a tangency involving an unstable manifold of dimension two are required. Homoclinic tangencies in this case were studied extensively in S. V. Gonchenko, V. S. Gonchenko, and Tatjer (2007) and Tatjer (2001). For a certain class of these tangencies, the so-called *Tatjer tangencies* in the sequel, a suitable renormalization and some subsequent simplifications allow to reduce the study of the dynamics in a neighborhood of the point of tangency to the study of the dynamics of the quadratic limit family given by

$$T_{\alpha,\beta}(x, y) = (\alpha + y^2, x + \beta y). \tag{2}$$

Numerical evidences in Pumariño and Tatjer (2007) show that the dynamical behavior of the above family is rather complicated. The attractors found for a large set of parameters seem to be two-dimensional strange attractors. Moreover, in Pumariño and Tatjer (2006) , a curve of parameters $(\alpha(t), \beta(t))$ was constructed in such a way that the respective map $T_{\alpha(t),\beta(t)}$ has an invariant region in $\mathbb{R}^2$ which is homeomorphic to a triangle. An analytical proof of the existence of two-dimensional strange attractors for $T_{\alpha,\beta}(x, y)$ in (2) is a formidable challenge. Up to now, the existence of these attractors have been proved for families of Expanding Baker Maps defined on a triangle, see Pumariño, Rodríguez, Tatjer, et al. (2014, 2015) and Pumariño, Rodríguez, and Vigil (2017, 2018, 2019). These piecewise linear maps are generalizations to $\mathbb{R}^2$ of the tent maps. It is remarkable to recall that the one-dimensional tent map gave a deep insight to prove of the existence of strange attractors for the limit family $f_a(x) = 1 - ax^2$ in (1) (see Benedicks and Carleson (1985) and Jakobson (1981)).

The brief review of results about diffeomorphisms that we have just furnished is basic to understand the contents of the first part of this book, but these results are still a step behind real-world applications. Models are usually families of differential equations, rather than families of diffeomorphisms. Therefore, a new question arises: How strange attractors appear in families of differential equations $\dot{u} = X_\mu(u)$?

A partial answer to the above question is well known for the flow of a vector field X in a neighborhood of a periodic orbit γ. Given a cross section Σ at a point $p \in \gamma \cap \Sigma$, the flow of X allows to define a first return map Π on Σ; this map is said the Poincaré map. Periodic points, strange attractors and, in general, invariant sets of f correspond to periodic orbits, strange attractors and invariant sets of X. This method is part of a more general process called suspension (not necessarily linked to a periodic orbit). See Katok and Hasselblatt (1995). Given a diffeomorphisms f defined on Σ is about looking for a vector field X whose flow defines f as a first return map on Σ. Then, one says that the vector field X was built by suspension from diffeomorphism f and the possible strange attractors of f correspond to strange attractors of X, which are said suspended strange attractors. Compare also with the Perturbation Principle of a Poincaré map in Pugh and Robinson (1983).

An immediate application of the Poincaré map is useful for non-autonomous differential equations of the type

$$\dot{u} = X(u) + Y(t, u, \mu) \quad u \in \mathbb{R}^2$$

where $Y(t, u, \mu)$ satisfies $Y(t, u, \mu_0) \equiv 0$ and it is a regular T-periodic perturbation of the Hamiltonian vector fields $X(u)$, which has a homoclinic loop.

Early research for these families was clearly driven by the pioneering work of Melnikov (1963), aimed at searching for transversal homoclinic orbits. Later, the ideas in Melnikov (1963), together with the exponential dichotomies (see appendices A and B) and the Lyapunov–Smith method, rounded off an analytical approach that was used in Chow, Hale, and Mallet-Paret (1980) to prove that if the Hamiltonian equation $\ddot{x} + q(x) = 0$ has a homoclinic orbit, then there must be a region $\mathcal{R}$ in the space of parameters (μ_1, μ_2), where the perturbed non-autonomous equation

$$\ddot{x} + q(x) = -\mu_1 \dot{x} + \mu_2 f(t)$$

also exhibits a transversal homoclinic orbit and periodic orbits with arbitrarily large period. In fact, these results are consistent with the well-known formation of horseshoes in a neighborhood of a transverse homoclinic point. Currently, after Mora and Viana (1993), we know that the formation or destruction of these horseshoes, and consequently of homoclinic tangencies, leads to the persistence of strange attractors (repellers). The region $\mathcal{R}$ is limited by two curves C_1 and C_2 which intersect at $(\mu_1, \mu_2) = (0, 0)$. Therefore, given a curve in the parameter space crossing either C_1 or C_2, strange attractors (repellers) will be persistent for the corresponding one-dimensional family.

The same analytical approach was applied in Costal and Rodríguez (1985) and Costal and Rodríguez (1988) to the case of two-parameter families of T-periodic three-dimensional vector fields

$$\dot{u} = X(u) + Y(t, u, \mu_1, \mu_2), \quad u \in \mathbb{R}^3. \tag{3}$$

Note that the Melnikov function had to be defined for a larger dimension. The conclusions are then like those in Chow, Hale, and Mallet-Paret (1980), but now the Poincaré map is a diffeomorphism in dimension three. Notice that here there is a proposal of dynamical models where two-dimensional persistent strange attractors may exist.

More examples focused on the dynamics close to a homoclinic or heteroclinic cycle can be found in the literature. An early example is the family of quadratic vector fields of Lorenz (1963). It exhibits a double homoclinic orbit to a saddle-node equilibrium point. However, we will be interested in the results that were given in the case of homoclinic orbits to a saddle-focus, that is, those involving a equilibrium point with complex eigenvalues. In case that the eigenvalues $-\alpha \pm i\omega$ and λ satisfy $\delta = \alpha/\lambda < 1$, we will refer to a Shilnikov homoclinic cycle. Dynamics associated with this type of cycles and its presence in generic unfoldings of low dimensional singularities are essential to illustrate the philosophy of the book.

Shilnikov homoclinic orbits are the simplest ones that yield infinitely many bifurcations and very complicated dynamics Shilnikov (1967). In fact, Shilnikov results are an extension of Birkhoff (1927) to a neighborhood of a homoclinic orbit Γ, where he proves the existence of a countable set of periodic solutions of saddle type. Again, a simpler geometric proof can be obtained by taking a return map Π on any local transverse section Σ at $p \in \Sigma \cap \Gamma$. According to Shilnikov (1970) and Tresser (1984), Π defines not only one but infinitely many linked horseshoe maps. In this way, it was shown that, for any $n \in \mathbb{N}$ and for any local transverse section Σ to the homoclinic cycle Γ, the return map Π has a compact invariant hyperbolic set on which it is topologically conjugate to the full shift on n symbols.

The following warning must be given: the return map Π is not a diffeomorphism in a neighborhood of $p \in \Sigma$ since, at least $\Pi(p)$ is not defined. However, this does not prevent us to be able to justify the presence of infinitely many horseshoes. In fact, Π is defined as the composition of two maps $\Phi : \Sigma_0 \to \Sigma_1$ and $\Psi : \Sigma_1 \to \Sigma_0$, where Σ_0 and Σ_1 are sections which are transverse to the vector field at p_0 and p_1, respectively. One of them, say Ψ, is a diffeomorphism, but the other, that is Φ, is defined by the flow near the equilibrium point but it is not defined at p_0. It is precisely Φ the map which, under the open condition $\delta < 1$, explains the arising of infinitely many horseshoes by the composition of both maps. When we consider a one-parameter family X_μ of vector fields such that X_0 has a Shilnikov homoclinic cycle, an infinite amount number of these horseshoes appear or disappear generically when $\mu \neq 0$ is close to 0. This mechanism of creation or destruction of horseshoes injects infinitely many homoclinic bifurcations for a sequence of parameter values μ_k which tends to $\mu = 0$ as k tends to ∞. Then, if the family X_μ is sufficiently regular one can apply the results in Mora and Viana (1993) to conclude that close to each μ_n there exist strange attractors (or repellers) for parameter values in positive Lebesgue measure set E_k. Attractors correspond to the dissipative case $1/2 < \delta < 1$.

Coexistence of strange attractors corresponds to intersections between the sets E_k. For this it is convenient that many horseshoes are created or destroyed simultaneously. This is easier to happen when the Shilnikov cycle remains, that is $\mu = 0$, but the sign of $\delta - 1$ changes. This was a motivation to study in Pumariño and Rodríguez (1997) saddle-focus homoclinic cycles with $\delta = 1$. Although this is a resonance condition, there are results of C^1 linearization which allow us to replace the map Φ by its linear part and to get an expression of Π which only depends on the diffeomorphism Ψ. Taking one of the coefficients of its linear terms as bifurcation parameter, a family Π_a is given. Associated with this family there is a sequence of invariant rectangles R_k. Restricted to R_k and after a convenient renormalization, the family Π_a transforms into a family $T_{a,b}$ whose limit family is

$$h_a(x) = x + \lambda^{-1} \log(a \cos x)$$

x

This is not the quadratic family, ideas and techniques in Benedicks and Carleson (1991) Mora and Viana (1993) were adapted in Pumariño and Rodríguez (1997) to prove the existence of strange attractors in each rectangle R_k with k arbitrarily large and for values of $a \in E_k$, with E_k a positive Lebesgue measure set. Later, it was proved in Pumariño and Rodríguez (2001) the existence of infinite sets E_k with a non-empty intersection and, therefore, the coexistence of infinitely many persistent strange attractors. The translation of these results to a family of vector fields was done in Pumariño and Rodríguez (1997) as well as in Pumariño and Rodríguez (2001) in the simplest possible way, that is, using piecewise regular vector fields. Results can be stated for C^3 families of vector fields which are linear in the neighborhood of the equilibrium point. Unlike Mora and Viana (1993), where Sternberg linearization results are required, only C^3 smoothness is needed in order to $T_{a,b}$ be a good C^3 unfolding of the family $\Psi_a(x, y) = (h_a(x), 0)$.

Inside the set of C^∞ families of vector fields with a saddle-focus homoclinic orbit verifying $\delta = 1$, those families that can be constructed from Pumariño and Rodríguez (2001) have simultaneously infinitely many persistent attractors, but they are not generic because the vector fields have to be linear in a neighborhood of the equilibrium point. In this same setting for $1/2 < \delta < 1$, using Colli (1998), Homburg (2002) proved the existence of generic families with infinitely many coexisting attractors, but the coexistence of these attractors is not a persistent property. The possible persistence of the coexistence of infinitely many strange attractors for these families of vector fields is, as far as we know, an open question.

Dynamics that we have described are all of them global in a neighborhood of the homoclinic cycle. The existence of these cycles is an essential ingredient. They are codimension one bifurcations which are difficult to detect in families of vector fields. Unlike transversal homoclinic tangencies that can be obtained for (3), homoclinic cycles for autonomous perturbations

$$\dot{u} = X(u) + Y(u, \mu_1, \mu_2), \quad u \in \mathbb{R}^3 \tag{4}$$

do not persist. Let X and Y be enough smooth vector fields with $Y(u, 0, 0) = Y(0, \mu_1, \mu_2) = 0$ and such that X has a homoclinic cycle. By means of the extension of the Melnikov function previously mentioned, the use of the exponential dichotomies and the Lyapunov–Schmidt reduction, it is proved in Rodríguez (1986) the persistence of the homoclinic cycle for parameter values on a curve $\mu_1 = \mu_1(s)$, $\mu_2 = \mu_2(s)$ with $\mu_1(0) = \mu_2(0) = 0$ and $\dot{\mu}_1(0)$ and $\dot{\mu}_2(0)$ well determined. Control on $\dot{\mu}_1(0)$ and $\dot{\mu}_2(0)$ and a suitable choice of the vector field X allow to define quadratic families in (4) with Shilnikov cycles. All these ideas were used to prove in Ibáñez and Rodríguez (1995) the presence of these cycles in generic unfoldings of a nilpotent singularity of codimension four in $\mathbb{R}^3$. Ten years later, a similar result was proved for the three-dimensional nilpotent singularity of codimension three in Ibáñez and Rodríguez (2005) with a different approach. By n-dimensional nilpotent singularity of codimension n we understand a vector field X such that $X(0) = 0$ and whose lineal part is linearly conjugate to

$$x_2 \frac{\partial}{\partial x_1} + x_3 \frac{\partial}{\partial x_2} + \cdots + x_n \frac{\partial}{\partial x_{n-1}}.$$

As follow from Dumortier and Ibáñez (1996), under generic assumptions, any unfolding of the three-dimensional nilpotent singularity of codimension three can be written as

$$y_2 \frac{\partial}{\partial y_1} + y_3 \frac{\partial}{\partial y_2} + \left(\nu_1 + \nu_2 y_2 + \nu_3 y_3 + y_1^2 + \varepsilon \kappa y_1 y_2 + O(\varepsilon^2)\right) \frac{\partial}{\partial y_3}. \quad (5)$$

where $(\nu_1, \nu_2, \nu_3) \in \mathbb{S}^2$, $(y_1, y_2, y_3) \in K \subset \mathbb{R}^3$, with and arbitrarily big compact set, $\kappa \neq 0$ and $\varepsilon \in \mathbb{R}^+$.

In Ibáñez and Rodríguez (2005) we prove that there exists a point $(\hat{\nu}_1, \hat{\nu}_2, \hat{\nu}_3) \in \mathbb{S}^2$ such that the corresponding vector field in the limit family obtained when $\varepsilon = 0$ exhibits a Bykov cycle, that is, a heteroclinic cycle consisting of two orbits connecting two saddle-focus equilibrium points with different stability index. In Barrientos, Ibáñez, and Rodríguez (2011) it is argued that family (5) is a generic unfolding of the cycle and, consequently (see details in Chapter 3), Shilnikov bifurcations are contained in the bifurcation diagram of the family. In Chapter 5 we include some details about the proof of the existence of the Bykov cycle. We conclude that strange attractors can emerge from codimension three singularities.

Since strange attractors only can appear for n-dimensional vector fields if $n \geqslant 3$ any attempt to reduce the codimension of the singularity must be restricted to those with a n-dimensional center manifold with $n \geqslant 3$. Nowadays it is clear that there exist codimension two singularities which generically unfold persistent strange attractors. Namely, as we will recall in Chapter 5, they can emerge from some Hopf–Zero singularities. There is a detail that makes a difference regarding genericity. All generic conditions which are required to guarantee that strange attractors appear in an unfolding of the three-dimensional nilpotent singularity of codimension three can be traced on a finite jet. On the contrary, the result about the existence of Shilnikov bifurcations in generic unfoldings of a Hopf–Zero singularity requires conditions which involve the full jet of the singularity.

Dimension three is the smallest dimension in which strange attractors can appear for flows of vector fields. These attractors are one-dimensional and, as we have already argued, Hopf–Zero singularities and nilpotent singularities become seeds for some of them. However, there exist attractors which can demand higher codimension singularities. For instance, the dynamics of the Lorenz attractor, which was proved to be strange by Tucker (2002), was previously explained by means of the geometric Lorenz attractor Guckenheimer and Williams (1979) and Williams (1979). In Dumortier, Kokubu, and Oka (1995) it was proved that a unfolding of the singularity $x_2\partial/\partial x_1 - x_1^3\partial/\partial x_2 + x_1^2\partial/\partial x_3$ contains geometric Lorenz attractors. The codimension of this singularity is very high. We do not know if the Lorenz attractor can be generically unfolded from a lower codimension singularity. A similar question could be posed for the spiral attractor, which according to the numerical simulations in Arneodo, Coullet, and Tresser (1981) (see also Pacifico, Rovella, and Viana (1998)), seems to exist in the neighborhood of a network consisting of two Shilnikov homoclinic orbits.

Two-dimensional strange attractors for vector field flows can only exist in dimension $n \geqslant 4$. These are suspended flows from three-dimensional diffeomorphisms that in turn possess such strange attractors. We have proposed as a mechanism for the formation of these two-dimensional strange attractors in families of diffeomorphisms the presence of certain class of Tatjer tangencies (involving a dissipative periodic point with two-dimensional unstable manifold). After an adequate renormalization in a neighborhood of the tangency point and some simplifications, the family of diffeomorphism becomes a good unfolding of the limit family (2). The dynamics that were numerically observed in Pumariño and Tatjer (2006) and Pumariño and Tatjer (2007) for this family are very promising. Strange attractors seem to exist, either simply connected, connected but not simply connected and not connected. For families of Expanding Baker Maps inspired by (2), see Pumariño, Rodríguez, Tatjer, et al. (2013) and Pumariño, Rodríguez, Tatjer, et al. (2014), the analytical proof of the existence of this type of strange attractors was possible. See Pumariño, Rodríguez, Tatjer, et al. (2015) and Pumariño, Rodríguez, and Vigil (2017, 2018, 2019) for the proof of the persistence of these strange attractors and the existence of a renormalization process that allows the successive duplication of the number of strange attractors. A challenging and hard process remains to be done in order to prove the persistence of strange attractors for family (2) and its unfoldings. We claim that this persistence is true.

We prove in Chapter 4 that for perturbations of vector fields that break a bifocus homoclinic cycle, under some extra assumptions, one could obtain Tatjer tangencies. A bifocus homoclinic cycle of a vector field in dimension $n = 4$ involves a homoclinic orbit to a bifocus equilibrium point, i.e., with two pairs of complex eigenvalues $-\alpha_1 \pm i\omega_1$ and $\alpha_2 \pm i\omega_2$ with $-\alpha_1 < 0 < \alpha_2$. However, the class of Tatjer tangencies that we will get do not involve a dissipative periodic point with two-dimensional unstable manifold. But, although this tangencies does not allow two-dimensional strange attractors, we can still obtain one-dimensional strange attractors and attracting invariant tori. Therefore, the neighborhood of a bifocus homoclinic orbit, despite being extremely rich and chaotic, does not seem to be the simplest scenario where two-dimensional strange attractors can appear. As we have anticipated, the proof of the existence of two-dimensional strange attractors on a flow and, additionally, the search for a singularity of low codimension from which these attractors can emerge, is a hard and challenging problem.

The dynamics near a bifocus homoclinic cycle was again studied by Shilnikov (1967, 1970) (see also Devaney (1976b), Fowler and Sparrow (1991), Ibáñez and Rodrigues (2015), and Wiggins (2013)), who proved the existence of a countable set of periodic solutions of saddle type. Once again, the existence of these periodic orbits is well understood from the existence of three-dimensional horseshoe maps. All this dynamics is found in generic unfoldings of the nilpotent singularity of codimension four in $\mathbb{R}^4$, once it was proved in Barrientos, Ibáñez, and Rodríguez (2011) that in such unfoldings there are bifocus homoclinic cycles. These cycles appear by continuation of similar homoclinic orbits present in the limit family of the nilpotent singularity which is a reversible Hamiltonian vector field. In Chapter 4 we will show that suspended robust heterodimensional cycles also appear by perturbation of conservative (Hamiltonian) and reversible bifocus homoclinic cycles.

Just now we can emphasize the interest of the study of the unfolding of singularities. Attention can be directed towards the study of the complexity that arises through the coupling of systems. For example, in cell growth the dynamics become more and more complicated in a space of very high dimension. A description of the dynamics in these cases is unattainable. A proposal, perhaps naive, could be to attend to the possible singularities that arise from the coupling provided that a catalogue of the new dynamics emerging from the new singularities is available. Restrictions on the model that come from structurally stable properties could make the problem manageable.

This book is written as notes for a course of the 32nd Brazilian Mathematics Colloquium. Therefore, it is not a compilation of many results that are known about chaotic dynamics for low-dimensional flows. These notes aim to acknowledge the relationship between certain singularities of low codimension and the existence of certain relevant dynamics, especially the persistence of strange attractors. In this way, both concepts, attractor and singularity, can evoke the singular attractors: those whose internal orbits are accumulated on a singularity. The canonical example of such attractors is the Lorenz attractor, already mentioned above. After this example, the singular attractors have been studied in several relevant articles Araujo et al. (2009), Bonatti, Pumariño, and Viana (1997), and Morales, Pacifico, and Pujals (1999, 2004), among others. For further details we address to Bonatti, Díaz, and Viana (2005, Ch. 9) where

xiv

relationship between the existence of singular attractors and the presence of cycles is explained.

This book consists of two parts. A first part will be mainly devoted to the study certain (homo)heteroclinic cycles (global bifurcations) of vector fields motivated by its presence in the second part of the book which deals with the unfolding of non-hyperbolic singularities (local bifurcations). Namely, after a revision in Chapter 1 of some result on local linearization around a equilibrium point, we study in Chapter 2 the dynamics near saddle-focus homoclinic cycles. Bykov cycles will also be dealt in Chapter 3 with mainly due to their role as organizing center in the formation of the saddle-focus homoclinic orbits for the unfolding of nilpotent singularities which are the focus of Chapter 5. Since both, generic unfoldings of nilpotent singularities and its limit families also present bifocus homoclinic cycles we dedicate Chapter 4 to study of this global bifurcations. The second part, Chapter 5, will be devoted to the study of the singularities of a vector field (local bifurcations). We will consider the nilpotent singularities of codimension n in $\mathbb{R}^n$, with $n = 3$ or $n = 4$ and we will prove that generic unfoldings of these singularities display Bykov cycles and Shilnikov homoclinic orbits if $n = 3$, and bifocus homoclinic orbits in the case $n = 4$. We will also study whether either the local codimension of the strange attractors in $\mathbb{R}^3$ is exactly three or a singularity of two codimension, the Hopf–Zero, can be taken as seed of strange attractors in $\mathbb{R}^3$. Results on dichotomy and continuation of (homo)heteroclinic orbits, which will be required for the exposition of this second part, will be collected in two appendices.

All the authors of this book were partially supported by the Spanish Research project MTM2017-87697-P. Alexandre Rodrigues was partially supported by CMUP (UID / MAT / 00144 / 2019), which is funded by FCT with national (MCTES) and European structural funds through the programs FEDER, under the partnership agreement PT2020. This author also acknowledges financial support from Program Investigador FCT (IF/00107/2015). Part of this work has been written during his stay in Nizhny Novgorod University, supported by the grant RNF 14-41-00044. Pablo G. Barrientos was partially supported by CNPq and CAPES and also thanks his wife Claudia A. Rivasplata for her assistance in figures' edition and the continuous support throughout the writing process of the book.

1

Local results of linearization

Consider a differential equation of the form

$$\dot{u} = X(u), \qquad u \in \mathbb{R}^n \tag{1.1}$$

where X is a sufficiently smooth vector field. In what follows $\varphi(t, u)$ denotes the associated flow that we will assume defined for every $t \in \mathbb{R}$. Each set $\gamma = \{\varphi(t, u) : t \in \mathbb{R}\}$ for $u \in \mathbb{R}^n$ is said to be a orbit of (1.1) and we will write $\gamma(t)$ to represent the parametrization of γ. A *heteroclinic cycle* of (1.1) is a finite collection of equilibria $\{O_1, \ldots, O_n\}$ together with a set of heteroclinic orbits $\{\gamma_1, \ldots, \gamma_n\}$ such that

$$\lim_{t \to -\infty} \gamma_j(t) = O_j \qquad \text{and} \qquad \lim_{t \to +\infty} \gamma_j(t) = O_{j+1}$$

for $j = 1, \ldots, n$ where $O_{n+1} \equiv O_1$. This means that the unstable manifold $W^u(O_j)$ intersects the stable manifold $W^s(O_{j+1})$. When $n = 1$, we say that the set $\Gamma = \{O_1\} \cup \gamma_1$ is a *homoclinic cycle*.

The destruction (or creation) of these cycles are called *heteroclinic or homoclinic bifurcations* and involve interesting dynamic transitions in a neighborhood of Γ. We refer to them as *global bifurcations* (they do not necessarily occur close to an equilibrium point). The goal of homoclinic and heteroclinic bifurcation theory is to investigate the recurrent dynamics near a homo/heteroclinic cycle. More precisely, we are interested in finding all orbits that stay in a fixed tubular neighborhood of a given homo/heteroclinic cycle Γ, for all time. A natural way

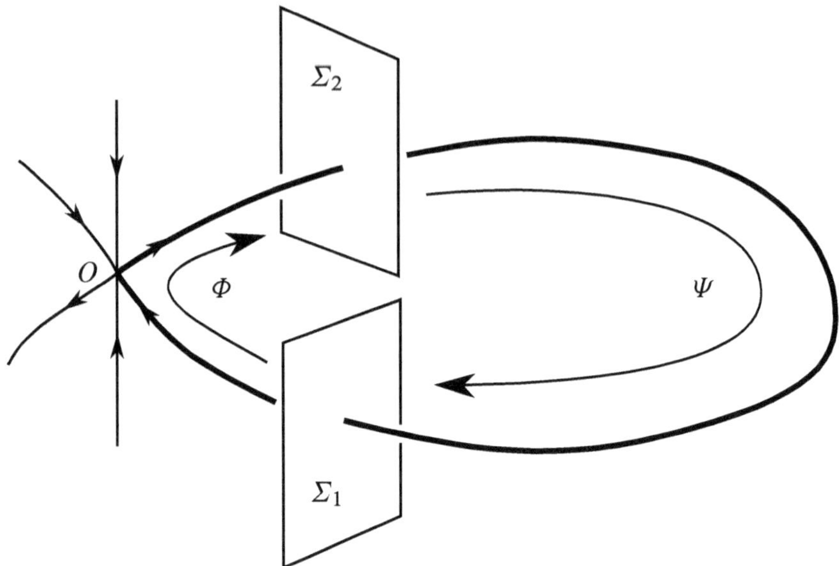

Figure 1.1: *In the classical approach, we consider two cross-sections Σ_1 and Σ_2 and write the first return map $\Pi = \Psi \circ \Phi$ which is the composition of transition maps between the cross sections: the local map Φ and global maps Ψ.*

for approaching this problem is to use *first-return maps* (or Poincaré maps), a technique we proceed to explain.

1.1   The Poincaré map

Denote by Σ a cross section to Γ. Suppose that $\Gamma \cap \Sigma = \{p\}$. Starting with an initial condition $u_0 \in \Sigma$, we then follow the solution $\varphi(t, u_0)$ until it hits Σ again, say at time $t = t_0$, and define the first-return map Π in such a way that

$$\Pi(u_0) = \varphi(t_0, u_0) \in \Sigma.$$

The iteration of Π reports on the dynamics of the vector fields near Γ: periodic points correspond to periodic orbits and invariant set, in general, correspond to invariant sets of the vector field. Even the asymptotic behavior of the orbits of X can be followed from the application Π. Moreover, attractors for X can be constructed by suspension from attractors of Π.

A remarkable difference arises between first return maps around a periodic orbit and a singular cycle Γ (a cycle which contains a singularity as homo/heteroclinic cycles). Since Γ contains some equilibrium point, hardly the Poincaré map Π is defined in a neighborhood of p because there is no finite time $t > 0$ such that $\varphi(t, p) \in \Sigma$. Actually, the closer an orbit to Γ is, the more time it spend near the equilibrium points of Γ, spoiling most finite time error estimates for Π. This difficulty is saved by considering Π as a composition of maps defined by means of cross sections Σ_i to Γ. Part of these maps are defined by the flow of the vector field near some equilibrium point O_j and they are said *local maps*. These local maps are the main factors in the dynamics of Π, while the rest of applications Π_i defined away from O_j are diffeomorphisms and they have less relevance. Certainly, except for questions related to the nature of the cycle (for instance in the double cycle Lorenz (1963)) the dynamics of Π and therefore that of the vector field close to Γ, depends only on the character of the equilibrium points.

The local return map

Consider two cross sections, say Σ_1 and Σ_2 that are, respectively, transverse to the local stable and local unstable manifold of a hyperbolic equilibrium point O that for simplicity we assume the origin in $\mathbb{R}^n$. The local transition map Φ is the first-return map from Σ_1 to Σ_2. Understanding Φ requires solving the linear differential equation

$$\dot{u} = DX(O)u \tag{1.2}$$

for $u \in \mathbb{R}^n$ close to O. The asymptotics of an orbit passing near O are, to a large extent, determined by the linearization (1.2) of (1.1) around the hyperbolic equilibrium point O. The closest eigenvalues to the imaginary axis will typically dominate the asymptotics as they give the slowest possible exponential rates: we therefore call them the *leading stable* and *unstable eigenvalues* of $DX(O)$. More precisely, denote the eigenvalues of $DX(O)$ by μ_i and ν_j for $i = 1, \ldots, s$, $j = 1, \ldots, u$ with $u + s = n$, repeated with multiplicity and ordered by increasing real part so that

$$\operatorname{Re}\mu_1 \leqslant \cdots \leqslant \operatorname{Re}\mu_s < 0 < \operatorname{Re}\nu_u \leqslant \cdots \leqslant \operatorname{Re}\nu_1. \tag{1.3}$$

The eigenvalues with $\operatorname{Re}\mu_s$ and $\operatorname{Re}\nu_u$ are the leading stable and unstable eigenvalues, respectively. The quotient between of the absolute value of the real part of the leading stable eigenvalue and the real part of the leading unstable eigenvalue, i.e.,

$$\delta = \frac{|\operatorname{Re}\mu_s|}{\operatorname{Re}\nu_u} > 0$$

is often referred to as the *saddle quantity* or *saddle index*.

The classical Hartman–Grobman Theorem states that there is a continuous coordinate change near the hyperbolic equilibrium O that transforms (1.1) into (1.2). Using such coordinates, the return map on a cross section is a homeomorphism but it is not clear how expansions and contractions in the new coordinates may be derived; they are essential for the bifurcation study. Assuming that the vector field is C^2, Belitskii (1973) derived an eigenvalue condition (sometimes called *non-resonant conditions*) which ensures the existence of a continuously differentiable coordinate change that linearizes the vector field near a hyperbolic equilibrium. Namely Belitskii proves the following result:

Theorem 1.1. *If X is a C^2 vector field and O is a hyperbolic equilibrium point of (1.1) with eigenvalues ordered as in (1.3) and such that for each $\rho \in \{\operatorname{Re}\mu_i : i = 1,\ldots,s\} \cup \{\operatorname{Re}\nu_j : j = 1,\ldots,u\}$ it holds*

$$\rho \neq \operatorname{Re}\mu_i + \operatorname{Re}\nu_j \quad \text{for all } i = 1,\ldots,s \text{ and } j = 1,\ldots,u.$$

Then there is a local coordinate transformation of class C^1 between (1.1) and (1.2).

Also C^1 linearization of (1.1) can be derived from the recent work of Newhouse (2017) for $C^{1+\alpha}$ vector fields, that is, for C^1 vector fields X with α-Hölder derivative DX for $0 < \alpha \leqslant 1$. closest

Theorem 1.2. *Let X be a $C^{1+\alpha}$ vector field and O a hyperbolic equilibrium point of (1.1) such that $DX(O)$ is bi-circular, that is, if the set consisting of the real parts of the eigenvalues of $DX(O)$ is a two element set $\{a,b\}$ with $a < 0 < b$. Then there is a local coordinate transformation of class C^1 between (1.1) and (1.2).*

In the particular, this result applies to C^r vector fields with $r > 1$ when O is a saddle-focus in dimension $n = 3$ or a bifocus in dimension $n = 4$ which are the framework that we are considering in this book. Also C^1 linearization can be followed from the papers of Samovol (1982, 1988) for three-dimensional flows. For higher-dimensional flows, under the assumption that leading eigenvalues are simple and unique (up to complex conjugation), the homoclinic center-manifold theorem Sandstede (1993) gives a three-dimensional homoclinic center manifold (if the homoclinic orbit is not in a specific form called flip configuration). This allows a geometric reduction to the three-dimensional case; although the homoclinic center manifold is, in general, only continuously differentiable with Holder continuous derivatives, we can still C^1 linearize (Shashkov and Turaev (1999)).

For higher regularity of the linearization, one must go to the works of Samovol (1979), Sell (1985), and Sternberg (1958) for sufficiently smooth vector fields. Namely, Sternberg's result ensures the following:

Theorem 1.3. *Let X be a C^r vector field and O a hyperbolic equilibrium point of (1.1) with eigenvalues denoted by λ_i, $i = 1, \ldots, n$ such that*

$$\lambda_i \neq \sum_{k=1}^{n} m_k \lambda_k \quad \text{for all non-negative integers } m_k \text{ with } \sum_{k=1}^{n} m_k > 1.$$

Then (1.2) is C^ℓ linearizing with $\ell = \ell(r)$. Moreover, $\ell(r) \to \infty$ as $r \to \infty$. In particular if $r = \infty$ then $\ell = \infty$.

Actually the above result follows by a similar result for diffeomorphisms applying the technique of Sternberg (1957) which implies that a C^r vector field has C^r linearization near the singularity O if and only if its time-one map has such a linearization at O. To be more specific the above result is a consequence of the following theorem for diffeomorphisms:

Theorem 1.4. *Let f be a C^r diffeomorphism defined in some neighborhood of the origin O of $\mathbb{R}^n$. Assume that O is a hyperbolic fixed point of f with eigenvalues λ_i for $i = 1, \ldots, n$ satisfying that*

$$\lambda_i \neq \lambda_1^{m_1} \cdots \lambda_n^{m_n} \quad \text{for all non-negative integers } m_k \text{ with } \sum_{k=1}^{n} m_k > 1.$$

Then f is C^ℓ linearizing near O with $\ell = \ell(r)$. Moreover, $\ell(r) \to \infty$ as $r \to \infty$. In particular if $r = \infty$ then $\ell = \infty$.

If the non-resonant condition is not met, the vector field may still be transformed into normal form (see for instance Chen (1963)). Structure preserving normal forms (e.g. in the class of equivariant, conservative or reversible differential equations) are another important topic: we will not discuss this here but refer to Bonckaert (1997) for Sternberg's theorem in a specific approach. Useful expansions for Φ may be hard to obtain if the normal form is not linear though problems involving one-dimensional unstable separatrices are often tractable.

Another approach used in Deng (1989b) is the use of *Shilnikov variables* introduced by Shilnikov in 1968 to compute the local transition map near equilibria to leading order. Instead of solving an initial value problem, solutions near the equilibrium are found using an appropriate boundary value problem. The analysis of the resulting integral formulae leads to asymptotic expansions for the solutions.

Saddle-focus
homoclinic
cycles

2

In three dimensions, a homoclinic orbit to a saddle-focus equilibrium point is the simplest cycle that yield infinitely many transitions and very complicated dynamics. At the mid 60's, under an eigenvalue condition that states that the expanding real eigenvalue dominates the complex conjugate contracting eigenvalues, Shilnikov proved analytically the existence of infinitely many periodic orbits of saddle type in each neighborhood of the homoclinic orbit Shilnikov (1965, 1967). This set of closed trajectories is similar to the extremely intrincated set of periodic orbits, mostly with a very long period, founded by Birkhoff (1927) near a homoclinic point of a diffeomorphism. In order to explain Birkhoff's result, Smale placed in Smale (1967) his geometrical device, the horseshoe, in a neighborhood of a transversal homoclinic point. Analogously, the periodic solutions found by Shilnikov are contained in suspended horseshoes that accumulate onto the homoclinic orbit as Tresser (1984) explained geometrically (see also Shilnikov (1970)). A deeply study of the unfolding of saddle-focus homoclinic orbits was done in Glendinning and Sparrow (1984) (see also Gaspard, Kapral, and Nicolis (1984)). They showed that there are sequences of parameters accumulating on the homoclinic bifurcation value at which the system has more geometrically complicated homoclinic bifurcations as well as a periodic orbit winds its way to infinite period. Dynamical features beyond hyperbolic suspended horseshoes and secondary homoclinic bifurcations, including the existence of periodic and strange attractors accumulating onto the homoclinic orbit, were described in later papers S. V. Gonchenko, Turaev, Gaspard, et al.

(1997), Homburg (2002), Ovsyannikov and Shilnikov (1987), and Pumariño and Rodríguez (1997, 2001).

In this chapter we will study the dynamics associated with the unfolding of saddle-focus homoclinic cycles describing the aforementioned results, among others, paying special attention to the existence of strange attractors.

2.1   Setting

Consider a differential equation

$$\dot{u} = X(u), \qquad u \in \mathbb{R}^3 \tag{2.1}$$

where X is a C^r vector field with $r > 1$. We assume that the associated flow satisfies the following hypotheses:

(S1) There is an equilibrium point O with eigenvalues of $DX(O)$,

$$\lambda \quad \text{and} \quad -\alpha \pm \omega i \quad \text{where } \lambda, \alpha > 0 \text{ and } \omega \neq 0.$$

(S2) There is (at least) one trajectory γ biasymptotic to O.

The homoclinic cycle is given by $\Gamma = \{O\} \cup \gamma$. Thus the equilibrium O possesses a two-dimensional stable manifold $W^s(O)$ and a one-dimensional unstable manifold $W^u(O)$ which intersect non-transversely. See Figure 2.1 for an illustration. If we restrict the system to the local stable manifold, the equilibrium O is a stable focus, i.e., the orbits spiral around O as $t \to \infty$. This is why O is called a *saddle-focus*.

We will consider different conditions on the *saddle quantity* (also called by *saddle index*) which is the absolute value of the quotient between of the real part of the contracting eigenvalue over the real part of the expanding eigenvalue, i.e.,

$$\delta = \frac{\alpha}{\lambda} > 0.$$

Namely, we study the above configuration under the assumptions:

(S3) $\delta < 1$ (Shilnikov condition)

(S4) $\delta > 1$

(S5) $\delta = 1$ (resonant condition).

A topological invariant or a modulus is a function of the vector field X that is invariant under topological equivalence. The saddle quantity δ, whether larger than one (condition **(S4)**) or not (condition **(S3)**), is always a topological invariant of saddle-focus homoclinic orbits Arnold et al. (1999), Ceballos and Labarca (1992), Palis (1978), Rodrigues (2015), and Togawa (1987). Also the absolute value of the frequency ω was also showed to be a topological invariant Dufraine (2001) for **(S1)** and **(S2)**.

Later, we will see that condition **(S4)** implies that Γ is attracting and the dynamics in a small tubular neighborhood of this homoclinic cycle is simple. Shilnikov discovered in Shilnikov (1965) that, when **(S3)** holds, the dynamics near the homoclinic solution Γ involves infinitely many periodic solutions arbitrarily close to the cycle. For this reason **(S3)** is usually called *Shilnikov condition*. In particular, we will refer us to Γ (resp. γ) as *Shilnikov homoclinic cycle* (resp. *orbit*) when **(S1)**–**(S3)** are satisfied. A geometric and nice explanation of the organization of these periodic orbits into infinitely many horseshoes has been given by Tresser (1984). Throughout the present chapter these results will be clarified.

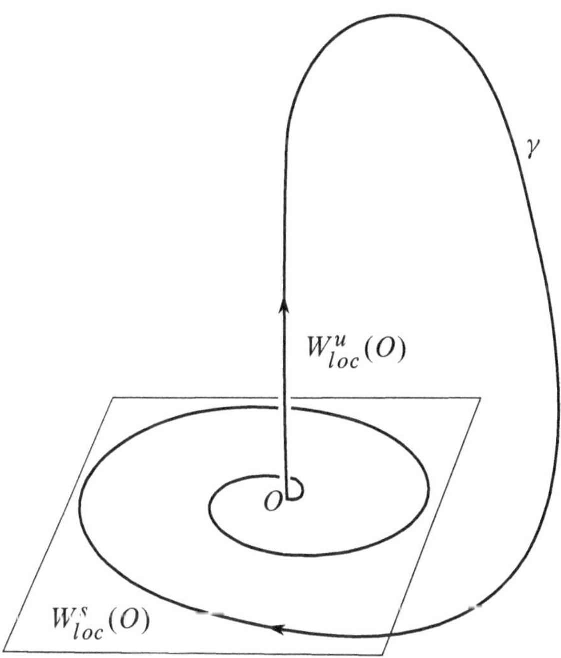

Figure 2.1: *Saddle-focus homoclinic cycle $\Gamma = \{O\} \cup \gamma$.*

In order to study the bifurcation theory of the above configuration, we consider a k-parameter family unfolding generically the homoclinic orbit γ in **(S2)**. In fact, it suffices to consider $k = 1$ since homoclinic orbits have codimension one (see in the appendix comments after Proposition A.21 and below Remark B.5). Thus, we will consider a *generic* one-parametric unfolding of (2.1) given by

$$\dot{u} = X_\mu(u) \stackrel{\text{def}}{=} X(u) + Y(u, \mu), \qquad u \in \mathbb{R}^3, \;\; \mu \in [-\varepsilon, \varepsilon] \qquad (2.2)$$

where Y is of class C^r with $r > 1$ such that $Y(0, \mu) = Y(u, 0) = 0$ and $\varepsilon > 0$ small enough. The word generic (in the unfolding) means that when $\mu \neq 0$, the stable and unstable manifolds of O split with non-zero speed with respect to the parameter. This system may be recast into the form

$$\begin{cases} \dot{x} = -\alpha x - \omega y + P(x, y, z, \mu) \\ \dot{y} = -\omega x + \alpha y + Q(x, y, z, \mu) \\ \dot{z} = \lambda z + R(x, y, z, \mu) \end{cases} \qquad (2.3)$$

where P, Q, R are C^r smooth along with their first derivatives vanishing at the origin for all μ and $\omega = \omega(\mu)$, $\alpha = \alpha(\mu)$ and $\lambda = \lambda(\mu)$ are positive. In this coordinates $(x, y, z) \in \mathbb{R}^3$, the local unstable manifold $W^u_{loc}(O)$ is tangent to the z-axis and the local stable manifold $W^u_{loc}(O)$ is a surface tangent to the plane defined by $z = 0$. In order to determine the nature of the dynamics near Γ we reduce its analysis to that of a Poincaré map on a small cross-section transversal to Γ.

2.2 First-return map

Since X is a C^r vector field with $r > 1$, according to Newhouse (2017) (see §1.1) we can C^1 linearize (2.3) around O. In the usual cylindrical coordinates (r, θ, z), the linearizations take the form

$$\begin{cases} \dot{r} = -\alpha r \\ \dot{\theta} = \omega \\ \dot{z} = \lambda z \end{cases} \qquad (2.4)$$

on a cylindrical neighborhood V of O of height $\varepsilon > 0$ and smaller radius. The local stable manifold $W^s_{loc}(O)$ of O (defined by $z = 0$ in this coordinates) divides V into two connected components, say $V^\pm$, well defined by the sign of the z-component. The boundary ∂V of V can be written as $\partial V = \texttt{In} \cup \texttt{Out}$ where $\texttt{In}$ is the cylinder wall and $\texttt{Out} = \texttt{Out}^+ \cup \texttt{Out}^-$ consists in the top and bottom caps of the cylinder distinguished by the sign of z. In the cylindrical coordinates $\mathtt{Out}^{\pm}$ corresponds to (r, θ) with $z = \pm\varepsilon$ respectively. Solutions starting at $V^{\pm} \setminus W_{loc}^{s}(O)$ leave in positive time the region of linear coordinates through $\mathtt{Out}^{\pm}$ respectively. We choose that the homoclinic orbit γ meets $\mathtt{Out}^{+}$. Hence the trajectories starting at V^{-} wanders away from the region that we are considering (a small neighborhood of Γ) and is of no further interest in this local analysis.

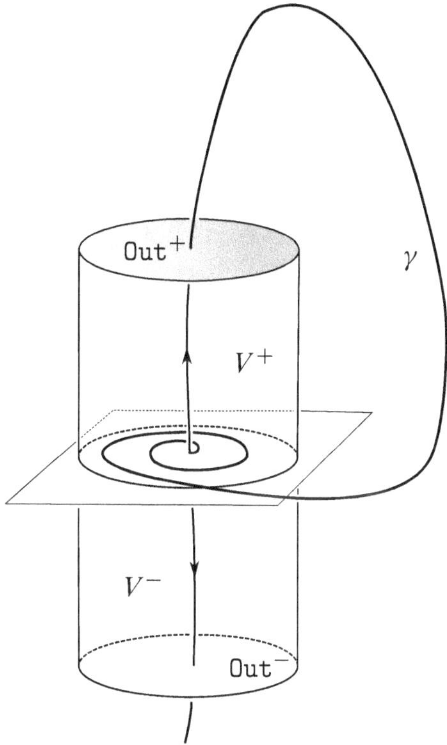

Figure 2.2: *Cylindrical neighborhood of O.*

The solution of (2.4) starting at a point (r, θ, z) in $V \setminus W_{\mathrm{loc}}^{s}(O)$ is given by

$$r(t) = re^{-\alpha t} \qquad \theta(t) = \omega t + \theta \qquad z(t) = ze^{\lambda t}.$$

The time of flight $T > 0$ for this solution to reach $\mathtt{Out}$ is found by solving the equation $\varepsilon = |z| \exp(\lambda T)$. Rescaling the variables, we may take $\varepsilon = 1$. Thus, this trajectory meets $\mathtt{Out}$ at

$$\bar{r} = r|z|^{\delta} \qquad \bar{\theta} = -\frac{\omega}{\lambda} \ln|z| + \theta \qquad \bar{z} = \pm\varepsilon \qquad (2.5)$$

where recall that $\delta = \alpha/\lambda > 0$.

Now, inside of V we choose a small two-dimensional rectangle Σ transverse to γ and also to $W^s_{loc}(O)$ intersecting it on a radial line out from O. Changing coordinates by a rotation around the z-axis if necessary, one can assume $\Sigma \subset \{\theta = 0\}$. That is, the coordinates on Σ are (r, z) with $\theta = 0$ and $r, |z|$ enough smaller. Hence, in rectangular coordinates

$$\Sigma \subset \{y = 0\}$$

i.e., Σ is in the xz-plane. Moreover, the radial coordinate r restricts to Σ coincides with the rectangular coordinate x. Thus, restricting the local map (2.5) to $\Sigma^+ = \Sigma \cap V^+$ we get that

$$(\bar{r}, \bar{\theta}) = \Phi(x, z) = \left(xz^\delta, -\frac{\omega}{\lambda} \ln z\right) \in \mathtt{Out}^+. \tag{2.6}$$

Local geometry.   Now, the main goal is to study the geometry of the local dynamics near the saddle-focus. First we introduce the concept of a *vertical and horizontal line.*

Definition 2.1. *A smooth curve β is called* vertical line *on Σ if β could be parameterized by a function $\beta : [0, 1] \to \Sigma$ such that $\beta(0) \in W^s_{loc}(O)$ and $\beta(s)$ has z-component monotonic and x-component bounded.*

The definition of vertical line may be relaxed: the components do not need to be monotonic for all $s \in [0, 1]$. We use the assumption of monotonicity to simplify the arguments.

Definition 2.2. *Let $a \in \mathbb{R}$, D be a disc centered at $p \in \mathbb{R}^2$.*

1. *A* spiral *on D around the point p is a smooth curve $\mathcal{S} : [a, \infty) \to D$ satisfying that $\mathcal{S}(s) \to p$ as $s \to \infty$ and such that if $\mathcal{S}(s) = (r(s), \theta(s))$ is its expression in polar coordinates around p then*

$$\lim_{s \to \infty} |\theta(s)| = \infty$$

   $r(s)$ is bounded by two monotonically decreasing maps converging to zero as $s \to \infty$ and $\theta(s)$ is monotonic for some unbounded subinterval.

2. *A* double spiral *on D around the point p is the union of two spirals accumulating on p and a curve connecting the other end points.*

3. *The region bounded by the* double spiral *on D around the point p is often called by* snake region *around p. Part of snake region is called a* curl.

Making use of (2.5), the next result characterizes the local dynamics near the saddle-focus, which is depicted in Figure 2.3.

Lemma 2.3. *A vertical line β on Σ is mapped by the local map into a spiral on* $\mathtt{Out}^{\pm}$ *accumulating on the point defined by* $\mathtt{Out}^{\pm} \cap W^u_{loc}(O)$.

Proof. Let β be a vertical line on Σ. Write $\beta(s) = (x(s), z(s))$, where $s \in [0, 1]$. Without loose of generality, we can assume that $z \geqslant 0$ is an increasing map as function of s and $z(0) = 0$. Using (2.6), the function Φ maps the vertical line β into the curve on $\mathtt{Out}^+$. This curve $\Phi \circ \beta$ is a spiral accumulating on the point defined by $\mathtt{Out}^+ \cap W^u_{loc}(O)$ because $z(s)$ converges monotonically to zero as $s \to 0$ and $x(s)$ is bounded. $\qquad\square$

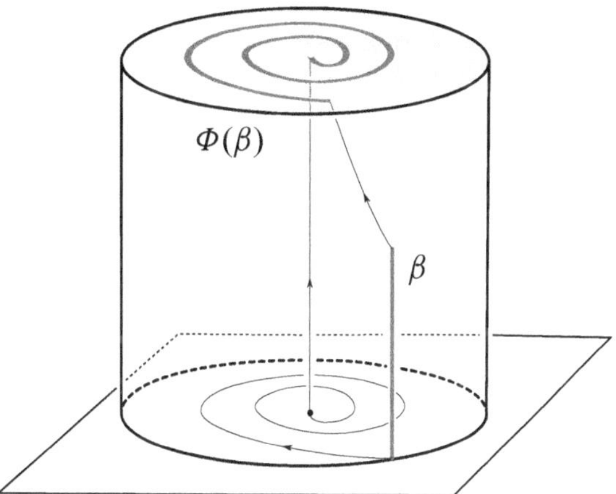

Figure 2.3: *Local dynamics near O: a vertical line β is mapped by Φ into a spiral on* $\mathtt{Out}$ *accumulating on the point defined by* $\mathtt{Out} \cap W^u_{loc}(O)$.

Global map. Recall that the cross-section Σ to Γ was chosen as a small rectangle contained in the plane $y = 0$. For $\mu \in [-\varepsilon, \varepsilon]$, the global map from $\mathtt{Out}^+$ to Σ, say Ψ, may be represented, by using regular coordinates on $\mathtt{Out}^+$

$$(x, y) = (r \cos \theta, r \sin \theta)$$

as the composition of a rotation of the coordinate axes and a change of scales. After a rotation and a uniform rescaling of the coordinates, we assume, without loss of generality, that for $\mu = 0$ the homoclinic orbit meets Σ at $(x, y, z) = (1, 0, 0)$. Moreover, since (2.3) is a generic unfolding when $\mu \neq 0$, the cycle Γ

is broken with non-zero velocity. Hence, on some open set $U \subset \mathtt{Out}^+$ contained the point $W^u_{loc}(O) \cap \mathtt{Out}^+$, the map Ψ may be written as

$$\begin{pmatrix} x \\ z \end{pmatrix} = \begin{pmatrix} 1 \\ \mu \end{pmatrix} + \begin{pmatrix} a_1 & a_2 \\ a_3 & a_4 \end{pmatrix} \begin{pmatrix} x \\ y \end{pmatrix} + \cdots \qquad (2.7)$$

where $a_i = a_i(\mu) \in \mathbb{R}$, the dots represents higher order terms and

$$\det \begin{pmatrix} a_1 & a_2 \\ a_3 & a_4 \end{pmatrix} \neq 0.$$

The first-return. Let us denote by $\Pi_\mu \overset{\text{def}}{=} \Psi \circ \Phi$ the *first-return map* from Σ^+ to Σ when it is well defined. Combining (2.6) and (2.7) the first-return take the form $(\bar{x}, \bar{z}) = \Pi_\mu(x, z)$ where

$$\begin{aligned}
\bar{x} &= 1 + a_1 xz^\delta \cos\left(\xi \ln z\right) + a_2 xz^\delta \sin\left(\xi \ln z\right) + \cdots \\
\bar{z} &= \mu + a_3 xz^\delta \cos\left(\xi \ln z\right) + a_4 xz^\delta \sin\left(\xi \ln z\right) + \cdots
\end{aligned} \qquad (2.8)$$

with $\xi = -\omega/\lambda$ and $\delta = \alpha/\lambda$. Up to high order terms, this first-return map may be approximated by

$$\bar{x} = 1 + axz^\delta \cos\left(\xi \ln z - \varphi_1\right) \qquad \bar{z} = \mu + bxz^\delta \cos\left(\xi \ln z - \varphi_2\right) \quad (2.9)$$

where $a = \sqrt{a_1^2 + a_2^2}, b = \sqrt{a_3^2 + a_4^2}$ and $\varphi_1 = \arctan a_2/a_1, \varphi_2 = \arctan a_3/a_4$.

2.3   Horseshoes

We will study Π_μ for $\mu = 0$. For simplicity of notation, we write Π instead of Π_0. The following lemma is illustrated in Figure 2.4.

Lemma 2.4. *The image of Σ^+ under Π is a region bounded by a double spiral accumulating on $\Gamma \cap \Sigma$ (snake region).*

Proof. Observe that the vertical boundaries of Σ^+ are vertical lines. Then, using Lemma 2.3, the image of Σ^+ on $\mathtt{Out}^+$ is a region bounded by a double spiral accumulating on $W^u_{loc}(O) \cap \mathtt{Out}^+$. Thus, $\Phi(\Sigma^+)$ is a snake region on $\mathtt{Out}^+$. Since the map Ψ is a (local) regular diffeomorphism by the flow box theorem, preserves the spiraling shape and then $\Pi(\Sigma^+)$ is contained in a region bounded by two spirals accumulating on $\Gamma \cap \Sigma$. $\qquad \square$

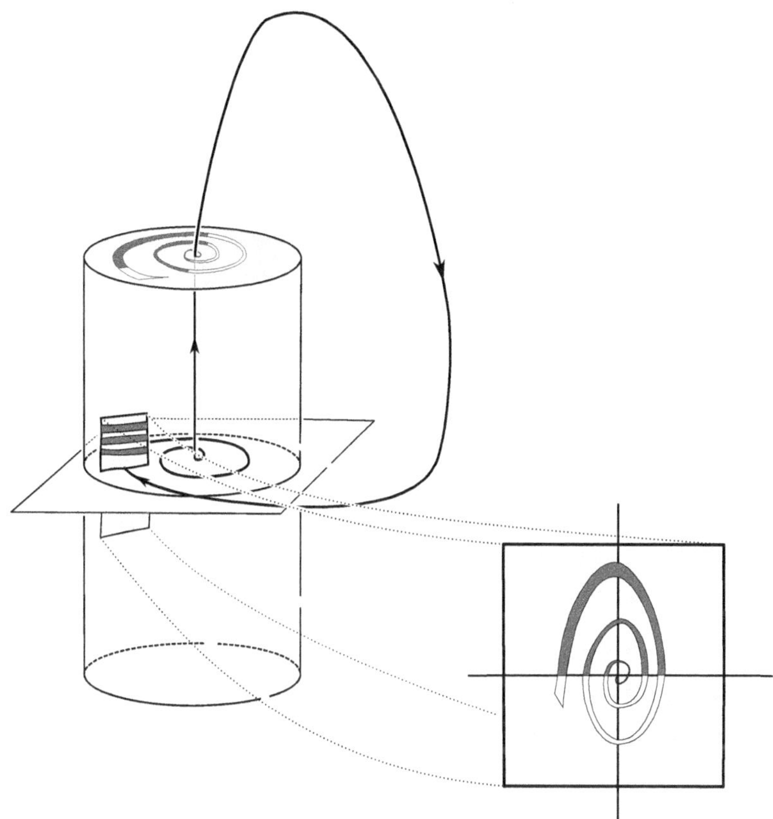

Figure 2.4: *The image of Σ^+ by Π is a region bounded by a double spiral accumulating on $\Gamma \cap \Sigma$ (snake region).*

The preimage of $W^s_{\text{loc}}(O) \cap \Sigma$ by the first-return map Π could be approximated by solving $\bar{z} = 0$ in (2.9). This provides an infinite sequence of lines given by

$$\ell_k : \qquad z_k(x) = e^{\frac{-\lambda k \pi}{\omega}}(1 + O(1)) \approx e^{\frac{-\lambda k \pi}{\omega}} \qquad (2.10)$$

for $k \in \mathbb{N}$ sufficiently large and x close to 1. As $k \to \infty$, these lines accumulate on $z = 0$. They are the intersection of the global manifold $W^s(O)$ with Σ^+. This is why the stable manifold is self-limiting here and has an helicoid form.

Let us select in the upper section Σ^+ the strips R_k bounded by the lines ℓ_k such that its image is still in Σ^+. By construction, the images of Σ^+ under Π which do not belong to the union of R_k fall into the region $\Sigma^- = \Sigma \cap V^-$ after some interaction and thus comes out of our analysis. Without lose of generality, we can assume the strip R_k in Σ^+ is bounded by the lines ℓ_{2k-1} and ℓ_{2k} and

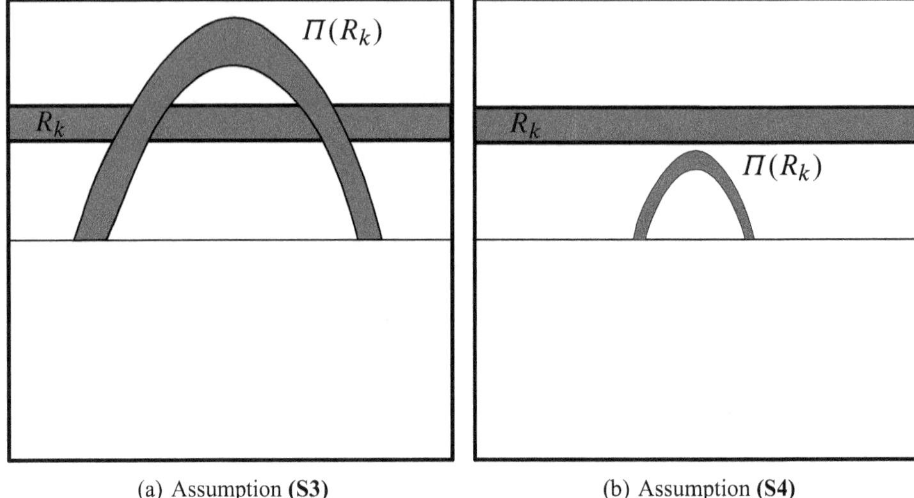

(a) Assumption **(S3)**   (b) Assumption **(S4)**

Figure 2.5: *(a) The horseshoe (b) Attracting structure of Γ.*

satisfies $\Pi(R_k) \subset \Sigma^+$. Having into account (2.10), each strip R_k may be approximated by the reparametrization

$$R_k \mapsto [1 - \tau, 1 + \tau] \times [z_{2k}, z_{2k-1}] \quad \text{where } z_i = e^{\frac{-i\pi\lambda}{\omega}}$$

and $\tau > 0$ is small enough. On the other hand, $\Pi(R_k) = \Psi \circ \Phi(R_k)$ is a curl, part of a bigger snake region (see Lemma 2.4). Note that the map Ψ adds only a bounded distortion factor to the picture induced by the local map Φ. Hence the image by Ψ of the half-curl $\Phi(R_k)$ is also a half-curl in Σ^+ of the same order. Therefore, $\Pi(R_k) = \Psi \circ \Phi(R_k)$ lies in an annulus in Σ centered at $(x, z) = (1, 0)$ and with inner and outer radius given, respectively, by

$$r_k^{\text{in}} \approx (z_{2k})^\delta \quad \text{and} \quad r_k^{\text{out}} \approx (z_{2k-1})^\delta. \tag{2.11}$$

The following result shows the existence of infinitely many suspended horseshoes in any neighborhood of Γ.

Theorem 2.5. *Let $\Pi : \Sigma^+ \to \Sigma$ be the first-return map of a vector field under the hypotheses (S1)–(S3). Then there is a sequence of hyperbolic compact invariant sets Λ_k in Σ^+ of Π such that the restriction of Π to Λ_k is topologically conjugate to a full shift on two symbols.*

Proof. The height of the top boundary of R_k may be estimated by z_{2k-1} while that the inner radius r_i^{in} of the curl $\Pi(R_k)$ is approximated by $(z_{2k})^\delta$. Since, by condition **(S3)**, $\delta < 1$ then we have that $2k - 1 > 2k\delta$ or equivalently $z_{2k-1} < (z_{2k})^{\delta}$ for any k large enough. Therefore, for each k large enough, the intersection $\Pi(R_k) \cap R_k \neq \emptyset$ and consists of two connected components of the intersection as Figure 2.5 shows. It is geometrically evident that this configuration provides a Smale horseshoe, i.e., a hyperbolic set topologically conjugate to the full shift of two symbols. $\qquad\square$

The following result shows that under the assumption **(S4)** instead of **(S3)** the dynamics near Γ is simple. Namely, the cycle attracts all trajectories starting in a small neighborhood $\mathcal{T}$ of Γ and remaining forward in this neighborhood.

Theorem 2.6. *Under the hypotheses **(S1)**, **(S2)** and **(S4)**, there are no periodic orbits in any small neighborhood $\mathcal{T}$ of Γ. Moreover, the cycle attracts the trajectories that remains forward in $\mathcal{T}$.*

Proof. Observe that, since $\delta > 1$ (condition **(S4)**), for every k large enough, $(2k - 1)\delta > 2k$ or, equivalently $(z_{2k-1})^{\delta} < z_{2k}$. This means as Figure 2.5 shows that the outer radius of the image of R_k by Π belongs below the bottom boundary of R_k and concludes the theorem. $\qquad\square$

Linked horseshoes. The set of all orbits that lie in a (fixed) small neighborhood $\mathcal{T}$ of Γ is larger, and much more complicated, than the union of the hyperbolic sets Λ_k in Theorem 2.5. Observe that each of these sets live in $\Pi(R_k) \cap R_k$. However, different R_k can interact and provide initial conditions of orbits jumping between different strips and providing new invariant dynamics in $\mathcal{T}$. See Figures 2.6. For a jump from R_i to R_j, with $i > j$, we need that

$$z_{2j-1} < r_i^{\text{in}} \approx (z_{2i})^{\delta} \qquad \text{or equivalently} \qquad 2j - 1 > 2i\delta. \qquad (2.12)$$

As a consequence since $\delta < 1$ we get that for every $n \geqslant 1$, we find a larger k such that $2k - 1 < 2(k+n)\delta$ and, consequently, $\Pi(R_i) \cap R_j$ has two connecting component for all $i, j \in \{k, \ldots, k + n\}$. Hence, the maximal invariant set in the union of R_i for $i = k, \ldots, k + n$ is a horseshoe in $N = 2n$ symbols, i.e., a hyperbolic invariant set of Π topologically conjugate to the full shift of $N = 2n$ symbols. Consequently one gets the following extension of Theorem 2.5.

Theorem 2.7. *Let $\Pi : \Sigma^+ \to \Sigma$ be the first-return map of a vector field under the hypotheses **(S1)**–**(S3)**. Then there is sequence of hyperbolic compact invariant sets Ω_k in Σ^+ of Π accumulating on $\Gamma \cap \Sigma$ such that the restriction of Π to Ω_k is topologically conjugate to a full shift on k symbols.*

Also the relation (2.12) suggests that the structure of the set of orbits that lie in $\mathcal{T}$ strongly depends on the saddle index $\delta < 1$ (hypothesis **(S3)**). Every orbit that remains in $\mathcal{T}$ in forward time has a natural coding a sequence of non-zero integers k_s. More precisely, fixed $\delta \leqslant v \leqslant 1$, one associates the symbols $+i$ or $-i$ to the two horizontal strips of the rectangle R_i whose image under Π cuts vertically the rectangles R_k for $k \geqslant vi$. Let $\Delta(v)$ be the set of such sequences $(k_s)_s$ in $(\mathbb{Z} \setminus \{0\})^{\mathbb{N}}$. The following theorem due to Shilnikov in Shilnikov (1970) (see also Tresser (1984)) describes the dynamics near Γ by means of subshift of finite type (a Markov chain) in infinite number of symbols:

Theorem 2.8. *Let $\Pi : \Sigma^+ \to \Sigma$ be the first-return map of a vector field under the hypotheses (S1)–(S3). Then for each v with $\delta \leqslant v \leqslant 1$ there is an invariant set $\Omega(v)$ in Σ^+ of Π such that the restriction of Π to $\Omega(v)$ is topologically conjugate to the shift map on $\Delta(v)$. Moreover, the restriction of Π to the intersection of $\Omega(v)$ with any compact subset in $\Sigma^+ \setminus W^s_{loc}(O)$ is hyperbolic.*

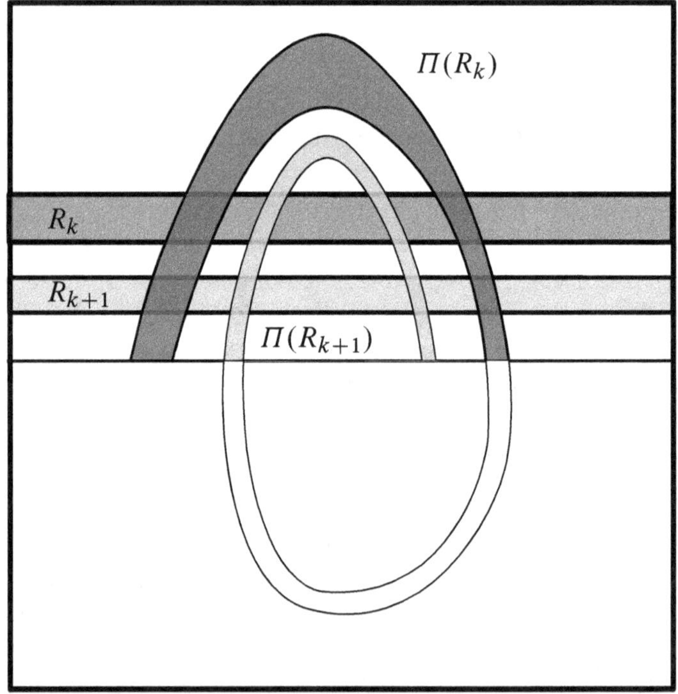

Figure 2.6: *Linked horseshoes.*

The dynamics described by Theorem 2.8 has several important properties. First, the suspension of $\Omega(\delta)$, i.e., as an invariant set of the system of differential equations, is not hyperbolic since its closure contains the equilibrium O. However, the restriction of this set to any compact set in $\Sigma^+ \setminus W^s_{loc}(O)$ is hyperbolic.

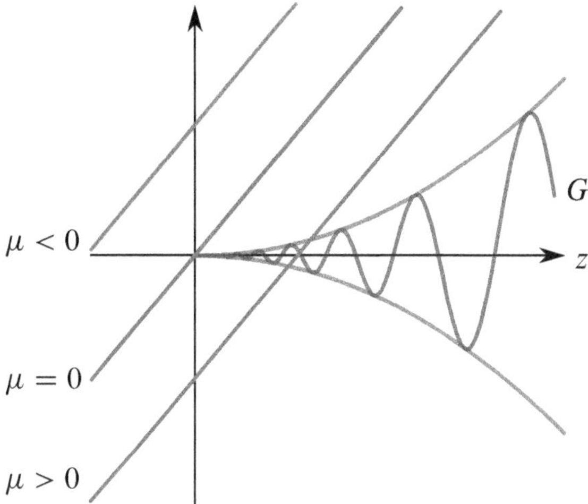

Figure 2.7: *Intersection of $G(z)$ with $\bar{z} = z - \mu$ under the assumption (S4).*

Secondly, the set $\Omega(\delta)$ does not, typically, admit a symbolic description with finitely many symbols. And, moreover, the set $\Omega(\delta)$ does not necessarily contain all solutions in the neighborhood $\mathcal{T}$. Note that $\Omega(\delta)$ depends sensitively on the $\delta < 1$. In particular, for any two, arbitrarily close saddle index δ_1 and δ_2, if $\delta_1 > \delta_2$, one can always find sequences in $\Delta(\delta_2)$ which are not in $\Delta(\delta_1)$. Thus, orbits with certain codings will disappear in $\mathcal{T}$ at an arbitrarily small increase of δ. This means bifurcations of non-hyperbolic periodic orbits and homoclinic tangencies occur within $\mathcal{T}$.

As mentioned above, for fixed compact set $\mathcal{K}$ in $\Sigma^+ \setminus W^s_{loc}(O)$, the set $\Omega(\delta) \cap \mathcal{K}$ is hyperbolic and thus persists for all systems which are sufficiently C^1-close, even when the homoclinic loop splits. The complete set $\Omega(\delta)$ does not entirely survive the breakdown of the loop; in fact, an infinitely many horseshoes close to Γ disappear as we see below.

2.4 Unfolding the homoclinic connection

Under the assumptions **(S1)**, **(S2)** and **(S4)** we have seen in Theorem 2.6 that for $\mu = 0$ the cycle Γ attracts all the orbits that remains forward in a small tubular neighborhood of Γ. Thus there is no periodic orbits near Γ. For $\mu \neq 0$, a local analysis due by Glendinning and Sparrow (1984) provides the following:

Theorem 2.9. *Let X_μ be a one-parameter family of vector fields unfolding at $\mu = 0$ the homoclinic cycle Γ of a vector field under the assumptions (S1), (S2)* *and (S4). Then an attracting periodic orbit appears for $\mu > 0$, collides Γ when $\mu = 0$ and disappears for $\mu < 0$.*

Proof. To study the fixed points of the first-return map, one needs to impose that $(x, z) = \Pi_\mu(x, z)$. Hence, imposing that $\bar{x} = x$, solving x as function of z and substituting into $\bar{z} = z$ in (2.9) gives

$$(z - \mu)(1 - az^\delta \cos(\xi \ln z - \varphi_1)) = bz^\delta \cos(\xi \ln z - \varphi_2). \qquad (2.13)$$

Solving (2.13) gives us the z-component of the fixed point. Substituting after that into (2.9) one gets the x-component of the fixed point. When z is small we can assume that $1 - az^\delta \cos(\xi \ln z - \varphi_1)$ is close to 1. Thus, the z-component of the fixed point will be modeled by the equation

$$z - \mu = bz^\delta \cos\left(\xi \ln z - \varphi_1\right).$$

Recall that the bifurcation parameter μ controls the splitting of the homoclinic loop Γ. Note that the original loop Γ exists at $\mu = 0$. Under the assumption **(S4)**, the graph of the one-dimensional map

$$G(z) = bz^\delta \cos\left(\xi \ln z - \varphi_1\right)$$

is shown in Figure 2.7. Variations of μ move the graph $\bar{z} = z - \mu$ up or down, so that an attracting fixed point appears for $\mu > 0$, collides with $z = 0$ for $\mu = 0$ and disappears for $\mu < 0$ concluding from this theorem. $\qquad\square$

As we have showed above, the dynamics for **(S1)**, **(S2)** and **(S4)** for $\mu \neq 0$ is still simple (the possible structures of the maximal invariant set is described in Shilnikov et al. (2001, Thm. 13.11)). Similar scenario can be found under the assumption that the equilibrium point has pure real eigenvalues (see Rodrigues (2017) and Shilnikov et al. (2001)). A major difference between these situations and the one described by **(S1)**–**(S3)** is that for Shilnikov homoclinic cycles, it is not necessary to break the connection to have complicated dynamics. Infinitely many horseshoes are always present in any tubular neighborhood of a Shilnikov homoclinic cycle and even strange attractor could be appears in its gaps (see §2.5). On the other, to study how these horseshoes are formed or destroyed when we break the connection we need to understand the unfolding of (2.8) for $\mu \neq 0$.

Assume now **(S1)**–**(S3)**. We know that the image of the strip R_k by Π_μ is also contained in an annulus with inner and outer radius given by r_k^{in} and r_k^{out} in (2.11) but now centered at $(x, z) = (1, \mu)$. Hence, in order to get a Smale horseshoe for Π_μ restricts to R_k for k large enough, it is necessary that

$$z_{2k-1} < \mu + r_k^{\text{in}} \approx \mu + (z_{2k})^\delta \quad \text{and} \quad z_{2k} > \mu + r_k^{\text{out}} \approx \mu + (z_{2k-1})^\delta.$$

As we have seen, both conditions hold for $\mu = 0$ when $\delta < 1$ and k large enough. However, if $\mu < 0$ there is $k_1 > 0$ such that $z_{2k-1} > \mu + (z_{2k})^\delta$ for all $k \geqslant k_1$. Similarly, if $\mu > 0$ we violated the second condition for all k large enough. Thus infinitely many horseshoes are destroyed. In particular only a finite number of periodic point of a given period survived for $\mu \neq 0$.

Subsidiary homoclinic orbits As we break the homoclinic orbit cycle Γ, also called as *principal homoclinic cycle*, other homoclinic orbits of a different nature could arise. Hence the Shilnikov scenario is repeated for these loops (see Figure 2.8). This phenomenon has been noted by Glendinning and Sparrow (1984). When Γ is broken up for $\mu > 0$, the set $W^u(O)$ intersects Σ^+ at the point $(x, z) = (1, \mu)$. Therefore, this point may be used as a new initial condition for the Poincaré return map. If the z-component of return of this point is zero, a new homoclinic orbit is found which pass twice through V before falling back into the origin.

Definition 2.10. *A n-pulse homoclinic orbit to O (with respect to neighborhood V of O) is a homoclinic solution that passes $n - 1$ times through V, before falling into O.*

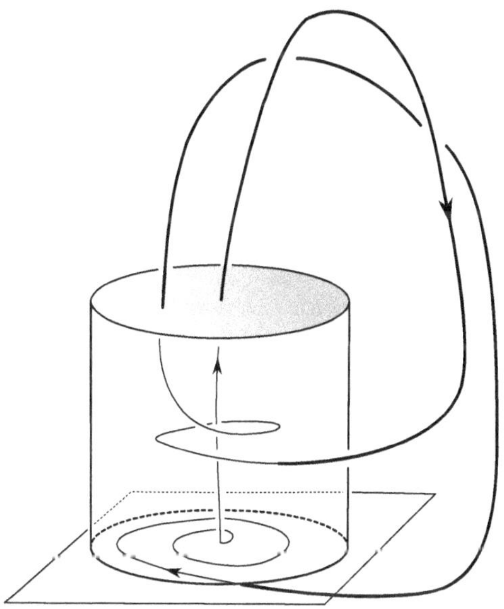

Figure 2.8: *Subsidiary homoclinic orbits.*

As suggested in Figure 2.8, we may state the following result:

Proposition 2.11 (Glendinning and Sparrow (1984)). *Let X_μ be a one-parameter family of vector fields unfolding at $\mu = 0$ the homoclinic cycle Γ of a vector field under the assumptions (S1)–(S2). Then, for any tubular neighborhood $\mathcal{T}$ of Γ,*

    (i) there are infinitely many values of μ converging to zero for which n-pulse homoclinic orbits provides (S3) holds;

    (ii) the only homoclinic is the principal cycle provides (S4) holds.

2.5   Strange attractors

Unless otherwise stated, in this section we will be considering a C^∞-family of vector fields X_μ satisfies the assumptions **(S1)** and **(S2)**. By C^∞-family we understand in both, in the dynamics and in the parameters. Actually, all the result stated holds if the family is C^r for some r sufficiently large.

In the case that Γ is a Shilnikov homoclinic cycle, that is, under the assumptions **(S1)–(S3)**, the existence of infinitely many suspended horseshoes was showed in Theorem 2.5 and 2.7. When these horseshoes are generated or destroyed the stable and unstable manifold of its periodic points inevitably present a homoclinic tangency. The unfolding of these homoclinic tangencies lead to complicate dynamics such as persistent strange attractors (or repellers). See definition below. Namely, we will study here the genesis of this attractors in two situations: in the unfolding of the homoclinic orbit ($\mu \neq 0$), i.e., when under perturbations there is no a homoclinic orbit, and when we still keep a Shilnikov homoclinic cycle but changing its internal configuration (by means of the parameter δ).

Definition 2.12. *A strange attractor is a compact invariant set Λ of a differentiable dynamics f having a dense orbit (transitivity) with at least one positive Lyapunov exponent (sensitive dependence on initial conditions), i.e., there is $z \in \Lambda$ such that $\{f^n(z) : n \geq 0\}$ is dense in Λ and*

$$\liminf_{n \to \infty} \frac{1}{n} \log \|Df^n(z)\| > 0$$

and whose stable set $W^s(\Lambda) = \{x : d(f^n(x), \Lambda) \to 0\}$ has non-empty interior. A strange repeller is strange attractor for f^{-1}. A vector field possesses a (suspended) strange attractor if a first return map to a cross section does.

Obviously, strange attractors are always non-trivial (i.e., non-periodic). However, they could be still hyperbolic as for instance the Plykin attractor or the

Smale solenoid. One of the first examples of a non-hyperbolic strange attractors was given (numerically) by Hénon (1976) for the family

$$H_{a,b}(x, y) = (1 - ax^2 + y, by)$$

at the parameters $a = 1.4$ and $b = 0.3$. This family for $b > 0$ can be easily written as

$$H_{a,b}(x, y) = (1 - ax^2 + \sqrt{b}y, \sqrt{b}x)$$

For $b = 0$ the dynamics of the corresponding Hénon map is given by the limit family of quadratic maps $f_a(x) = 1 - ax^2$. The compact interval $[1 - a, 1]$ for $1 < a \leqslant 2$ is an invariant set of f_a. But also of the so-called tent map $h_a(x) = 1 - a|x|$. For any parameter a, it is easy to find dense orbits of h_a which are clearly expansive. However this is not true for the family f_a. Mainly because the derivatives are arbitrarily small near 0, there appear sinks for many values of the parameter a. For other values it was proved in Benedicks and Carleson (1985) that there exists $0 < \varepsilon \ll 1$ and a positive Lebesgue measure set $E \subset (2 - \varepsilon, 2)$ such that the orbit $\{f_a^n(0) : n \geqslant 0\}$ is expansive and dense in $[1 - a, 1]$ for every $a \in E$. Hence $[1 - a, 1]$ is a strange attractor of f_a for every $a \in E$. The results in Benedicks and Carleson (1985), see also Jakobson (1981), were key to prove the existence of strange attractors for the family of Hénon maps in Benedicks and Carleson (1991). For $b > 0$, the maps $H_{a,b}$ are diffeomorphisms with a saddle fixed point p. When b goes to zero the unstable manifold $W^u(p)$ of p approaches the segment $\ell = \{(x, 0) \in \mathbb{R}^2 : x \in [1 - a, 1]\}$ and it is natural to think that the dynamics of $H_{a,b}$ along $W^u(p)$ are close to the dynamics of f_a. From this idea, a network of complicated arguments allow to check that for b arbitrarily small ($0 < b \ll 1$) and $a \in E$ (where E is again a positive Lebesgue measure set close to $a = 2$) there exists an expansive orbit that is dense in the closure of $W^u(p)$ and consequently, the closure of $W^u(p)$ is a strange attractor of $H_{a,b}$. After a careful reading of the intricate arguments in Benedicks and Carleson (1991) it is natural to think that an extension of its conclusions could be possible for some other family $F_{a,b}$ whose family of limit maps was also the quadratic family f_a. Families of this type will be called *Hénon-like families*:

This family can be include in the following definition:

Definition 2.13. *A C^r-family $(g_a)_a$ of diffeomorphisms with $r \geqslant 3$ is said to be Hénon-like family if g is C^3-close enough to the family $(\Phi_a)_a$ of endomorphisms of $\mathbb{R}^2$ given by*

$$\Phi_a(x, y) = (1 - ax^2, 0)$$

on a sufficiently large rectangle in $\mathbb{R} \times \mathbb{R}^2$ (say, $[-4, 4] \times [-10, 10]^2$).

Hénon-like maps are (strongly) area-contracting everywhere in their domain and so they can not have non-trivial hyperbolic attractors. This lack of hyperbolicity prevents to find attractors which persist under perturbations. Hence, the question that arises is whether such attractors could survive not for all perturbations but, at least, for most. Since, it is not known how to introduce a notion of measure for the space of dynamics (at least in dimension large than one[1]), Kolmogorov in his plenary talk ending the ICM 1954 proposed to consider finite parametric families (actually, finite number of functional, see Hunt and Kaloshin (2010)).

Definition 2.14. *Let $\mathcal{P}$ be a property of a dynamics. A parametric family $(f_\mu)_\mu$ exhibits* persistently *the property $\mathcal{P}$ if $\mathcal{P}$ is observed for f_μ in a set of parameter values μ with positive Lebesgue measure. The property $\mathcal{P}$ is said to be* typical *(in the sense of Kolmogorov) if there is a Baire generic set of parameter families of dynamics exhibiting the property $\mathcal{P}$ persistently.*

Mora and Viana (1993) proved that any Hénon-like family $(g_\mu)_\mu$ has a set of parameters with positive Lebesgue measure for which g_μ has a strange attractor. Thus, strange attractors coming from a Hénon-like family, which are called *Hénon-like strange attractors*, are always non-hyperbolic and persistent. Moreover, attractors appear typically (in the sense of Kolmogorov) in the unfolding of two-dimensional homoclinic tangencies as we will explain below.

Generic unfoldings of the homoclinic connection.   First of all, we will deal with the case of generic unfoldings of a Shilnikov homoclinic orbit as in system (2.2). That is, when the stable and unstable manifold of O split with non-zero velocity with respect to the parameter μ. As mentioned, there exist infinitely many horseshoes Λ_k for the first-return map Π_μ for the value of the parameter $\mu = 0$. If $\mu \neq 0$ only a finite number of these horseshoe survive. In particular, as μ grows from a negative value to $\mu = 0$ an infinite number of horseshoes is generated and then after μ goes through $\mu = 0$ these horseshoes should be destroyed. Thus, for many parameters μ_0 arbitrarily close to $\mu = 0$ we find a periodic point p in one of the horseshoe of Π_{μ_0} so that the stable and unstable manifold of p have an homoclinic tangency which is in turn broken when the parameter μ varies. Then the family Π_μ is an unfolding of the homoclinic tangency at $\mu = \mu_0$ and one can apply the following theorem from Mora and Viana (1993).

Theorem 2.15. *Let $(f_\mu)_\mu$ be a C^∞ one-parameter family of diffeomorphisms on a surface and suppose that f_{μ_0} has a homoclinic tangency associated with*

---

[1] For one-dimensional dynamics the Malliavin–Shavgulidze measures has been proposed as a good analogous to Lebesgue measure. See Triestino (2014)

some periodic point p. Then under generic (even open and dense) assumptions, there is a positive Lebesgue measure set E of parameter values near μ_0 such that for $\mu \in E$ the diffeomorphism f_μ exhibits a Hénon-like strange attractor, or repeller, near the orbit of tangency.

Another similar result, usually called as *Newhouse phenomenon* is the co-existence of infinitely many hyperbolic periodic sinks (or sources) for a local residual set of C^2 diffeomorphisms containing the diffeomorphism displaying the tangency in the closure Newhouse (1979). Based in Newhouse original work, Robinson (1983) proved the following parametric version of the Newhouse phenomenon:

Theorem 2.16. *Let $(f_\mu)_\mu$ be a C^∞ one-parameter family of diffeomorphisms on a surface and suppose that f_{μ_0} has a homoclinic tangency associated with some periodic point p. Then under generic (even open and dense) assumptions, there are non-trivial intervals I arbitrarily near $\mu = \mu_0$ such that for a residual set of parameters μ in I, the diffeomorphism f_μ has infinitely many hyperbolic periodic sinks (or sources) near the orbit of tangency.*

Although the above theorems are stated for surface diffeomorphisms they are local results and thus could be applied to the Poincaré map associated with X_μ. The proof of both results need a renormalization process that is performed from a C^2 linearization of f_μ near p_μ where p_μ is the continuation of the periodic point p_{μ_0} involved with the homoclinic tangency. For this reason, f_μ needs to be sufficiently smooth in both the dynamics and the parameters. Namely, according to Sternberg (1958) C^2 linearizing μ-dependent coordinates exist provided that the family f_μ is sufficient regular and some generic (even open and dense) non-resonant conditions are satisfied for the eigenvalues of f_μ at p_μ (c.f. §1.1 and see Palis and Takens (1993, pg. 47)). Actually, all this process could be performance by C^r families with r large enough (possibly $r \geq 3$ as stated in Robinson (1983)).

On the other hand, observe that to prove the existence of infinitely many horseshoes in a neighborhood of the Shilnikov homoclinic cycle Γ we have considered C^1 linearization of X_μ near O. Hence the first-return map that we have obtained in this linearizing coordinates (i.e., (2.8)) is only C^1 conjugate to the Poincaré map associated with X_μ. Actually we have proved the existence of horseshoes using (2.9) which is only a C^1 approximation of this first-return map in linearizing coordinates. But this C^1 approximation and C^1 conjugation are enough to conclude that the Poincaré map associated with X_μ, which is as regular as the vector field, have infinitely many horseshoes for $\mu = 0$ and finitely many for $\mu \neq 0$. Therefore, since we are considering a C^∞ family X_μ in order to avoid the above mentioned difficulties (C^r families with $r \geq 3$

maybe it suffices) applying Theorem 2.15 and Theorem 2.16 to this Poincaré map we get the following:

Theorem 2.17. *Let $(X_\mu)_\mu$ be a C^∞ one-parameter family of three-dimensional vector fields and assume that X_0 satisfies (S1)–(S3). Then under generic (even open and dense) assumptions, for every neighborhood $\mathcal{T}$ of the homoclinic cycle Γ it holds that*

> (i) *there are non-trivial intervals I of parameters arbitrarily near $\mu = 0$ such that for a residual set of parameters $\mu \in I$, the vector field X_μ has infinitely many hyperbolic periodic attractors, or repeller, in $\mathcal{T}$;*

> (ii) *there is a set E of parameters with positive Lebesgue measure and arbitrarily near $\mu = 0$ such that for every $\mu \in E$, the vector field X_μ has suspended Hénon-like strange attractors, or repeller, in $\mathcal{T}$.*

Another comment must also be done regarding the presence of attractors or repellers, according to the determinant of $D\Pi_\mu$ less than one or greater than one. It is not difficult to see that the determinant of $D\Pi_\mu$ at a point $p = (x_\star, z_\star)$ in some of the horseshoe Λ_k only contains terms of order $z_\star^{2\delta-1}$. Hence we need to distinguish two sub-types of the hypothesis **(S3)**:

(S3a) $0 < \delta < 1/2$ (expansiveness)

(S3b) $1/2 < \delta < 1$ (dissipativeness).

The map Π_μ will be area expanding (hypothesis **(S3a)**) or area contracting (hypothesis **(S3b)**) for $z_\star$ sufficiently small. Consequently, we get persistent strange attractors for **(S3b)** and persistent strange repeller for **(S3a)**.

Non-generic unfoldings of the homoclinic connection. Denote by $\mathcal{S}_1$ the set of C^∞ vector fields X such that the system (2.1) satisfies the condition **(S1)**, **(S2)** and **(S5)**. Observe that $\mathcal{S}_1$ is the intersection of two codimension one submanifolds in the open set of C^∞ vector fields X satisfying **(S1)**. Namely the submanifolds of vector fields satisfying **(S2)** and **(S5)** respectively. Thus, $\mathcal{S}_1$ is a codimension two submanifold of three-dimensional smooth vector fields. For a regular two-parameter families X_ν that are transversal to $\mathcal{S}_1$ at $\nu = 0$ one gets a curve $\nu(a)$ in the two-parameter space emanating from $\nu = 0$ so that on this curve the family $X_{\nu(a)}$ has a Shilnikov homoclinic cycle Γ_a (see Remark B.5 in Appendix B). For each a, the two-parameters family X_ν could be seem as a generic unfolding of Γ_a breaking the connection. Thus, as before, since infinitely many horseshoe are created and destroyed, infinitely many homoclinic bifurcations takes place for a sequence of parameters ν_n going to $\nu = 0$. Each of these bifurcations is the seed of persistent Hénon-like strange attractors like those discussed above. Its persistence is proven by means of a renormalization that only guarantees the infinitesimal size of the attractors. For $v = 0$ these attractors could simply collapse. It seems, then, justified to study the possible persistence of strange attractors for families X_a in $\mathcal{S}_1$.

In view of this, we will focus on non-generic unfoldings of the homoclinic connection of a vector field in $\mathcal{S}_1$. Namely we will consider one-parameter families X_a having a saddle-focus homoclinic cycle Γ with $\delta(a) = 1$ for each a in a real interval I. That is, such that $X_a \in \mathcal{S}_1$ for all $a \in I$. In this framework, one has to notify that no value of the parameter verifies the non-resonance conditions in Sternberg (1958) (see §1.1) and, therefore, one cannot be sure the existence of a C^r linearization with $r > 1$. This extra regularity would be needed to carry on (by the conjugation) the results on the existence of persistent strange attractors from the first-return map in linearizing coordinates to the Poincaré map associated with $X_a \in \mathcal{S}_1$. Thus, in order to avoid this difficulty, Pumariño and Rodríguez in Pumariño and Rodríguez (1997) and Pumariño and Rodríguez (2001) consider special one-parameter C^∞ families of vector fields X_a in $\mathcal{S}_1$ defined by joining smoothly (by a C^∞-bump function) a linear field in a given neighborhood of origin with a family of fields defined by suspension from a one-parameter family of linear diffeomorphisms between two cross sections to cycle Γ. Namely, in Pumariño and Rodríguez (1997, 2001) the authors prove the following result.

Theorem 2.18. *There is a one-parameter family of vector fields X_a in $\mathcal{S}_1$ such that for every neighborhood $\mathcal{T}$ of the homoclinic cycle Γ it holds that*

(i) there is an open set P of parameter such that for every $a \in P$ the vector field X_a has infinitely many periodic attractors in $\mathcal{T}$;

(ii) there is a set E of parameters with positive Lebesgue measure such that for every $a \in E$ the vector field X_a has infinitely many strange attractors.

We must notify that actually the proof of the above result not need a higher class than C^3. The family X_a in the previous statement is chosen linear in the neighborhood V of the origin so that the expression for the local map Φ is the same as (2.6) with $\delta = 1$ and $\omega = -1$. Outside of V, if no restriction is made, Ψ is given again by (2.7) with $\mu = 0$. Proceeding in this way, four parameters a_i already appear. These are four parameters that, unlike what happened in the previous cases, will now play an essential role. To have a one-parameter family X_a as simple as possible the vector fields out of the neighborhood V are taken, by means of a suspension process, in such a way that $a_1 = a_4 = 0$, $a_2 = 1, a_3 = a$ and the non-linear terms disappear from (2.7). Thus, the family of first-return map resulting in (2.8) is

$$\Pi_{\lambda,a}(x, z) = (1 + xz \sin(\lambda^{-1} \log z), axz \cos(\lambda^{-1} \log z)). \qquad (2.14)$$

In order to prove that X_a has infinitely many strange attractor it is sufficient to prove that $\Pi_{\lambda,a}$ does. With this in mind, we will introduce some changes of variables which will allows us to prove that the restriction of $\Pi_{\lambda,a}$ to the rectangle R_k given in §2.3 for k large enough is C^3 close to family $F_{\lambda,a}(u, v) = (f_{\lambda,a}(u), 0)$ of endomorphisms of $\mathbb{R}^2$ where $f_{\lambda,a}$ is the one-dimensional map

$$f_{\lambda,a}(u) = \lambda^{-1} \log a + u + \lambda^{-1} \log \cos u. \qquad (2.15)$$

By means of new coordinates $u = \lambda^{-1} \log z + 2\pi k$ and $v = b^{-1/2}(x - 1)$ with $b = \exp(-2\pi\lambda k)$ for some k large enough (2.14) becomes in

$$\Pi_{\lambda,a,b}(u, v) = (f_{\lambda,a}(u) + \lambda^{-1} \log(1 + \sqrt{b}\, v), \sqrt{b}(1 + \sqrt{b}\, v)e^{\lambda u} \sin u).$$

The maps $\Pi_{\lambda,a,b}$ has a forward invariant set given by

$$U_k = \left\{ (u, v) : u \in I, -\sqrt{b}\, e^{(\ell - \frac{\pi}{2})\lambda} \leqslant v \leqslant \sqrt{b}\, e^{(\ell + \frac{\pi}{2})\lambda} \right\}$$

which is contained in the rectangle R_k in the new variables and where $I \subset (-\pi/2, \pi/2)$ is a forward invariant interval for the one-dimensional map (2.15) and ℓ is the smallest of the lengths of the connected components of $I \setminus f_{\lambda,a}(I)$. In Pumariño and Rodríguez (1997), it is proved that the restriction of $\Pi_{\lambda,a,b}$ to U_k for k large enough exhibits persistent strange attractors. That is, there exists a set E_k of positive Lebesgue measure such that for every $a \in E_k$ the map $\Pi_{\mu,a,b}$ has a strange attractor contained in U_k. This was obtained by proving that the family $\Pi_{\lambda,a,b}$ is a *good* unfolding of the two-dimensional map

$$F_{\lambda,a,b}(u, v) = (f_{\lambda,a}(u), 0)$$

and that the unimodal family $f_{\lambda,a} : I \to I$ has similar properties to the quadratic family. A good unfolding provides that the first return map $\Pi_{\lambda,a,b}$ is C^3 close to the family $F_{\lambda,a}$ for b sufficiently small. On the other hand, the similarity of $f_{\lambda,a}$ with the quadratic family allows to conclude the persistence of strange attractors for $f_{\lambda,a}$.

The jump from the unimodal family $f_{\lambda,a}$ to the two-dimensional first-return map $\Pi_{\lambda,a,b}$ is long and laborious. It requires a careful transfer to this scenario the scheme developed in Benedicks and Carleson (1991) and Mora and Viana (1993) for the Hénon and Hénon-like families, respectively. All this to get an expansive orbit that will turn out to be dense in the unstable manifold of a saddle fixed point of $\Pi_{\lambda,a,b}$. Once proved that for each $a \in E_k$ there was a strange attractor in U_k the persistence of the coexistence of infinitely many strange attractors follows by proving that there are infinitely many sets E_k whose intersection has a positive Lebesgue measure. In fact, the goal of Theorem 2.18 and, firstly, of the book Pumariño and Rodríguez (1997) was in its day the challenge of understanding the hard task developed first by Benedicks and Carleson (1991) and later by Mora and Viana (1993) and applying their ideas to the case of families of vector fields in $\mathcal{S}_1$.

As mentioned, when in a two parameter family X_ν unfolding generically a vector field in $\mathcal{S}_1$ for $\nu = 0$, infinitesimal persistent Hénon-like strange attractors could be obtained which could simply collapse for $\nu = 0$. In addressing this problem, two questions emerged: one good and the other not. Regarding the first, a renormalization was possible in rectangles U_k whose size depended on $f_{\lambda,a}$ and it did not collapse (possible non-infinitesimal attractors). Regarding the second, the family of return maps $\Pi_{\lambda,a,b}$ was not now a Hénon-like family: the limit family $f_{\lambda,a}$ is quite different from the quadratic family that appears as a limit family in the Hénon-like case. In order to overcome this difficulty, a thorough review of Benedicks and Carleson (1991) and Mora and Viana (1993) was necessary to verify that all the ideas in these references could be applied to the new case. This work offers the conviction of a certain universality in the methodology of Benedicks and Carleson (1991), which can be found in the later papers by Wang and Young (2001, 2008).

In the time that elapsed between the writing of Pumariño and Rodríguez (1997) and Pumariño and Rodríguez (2001), Colli (1998) proved the following result:

Theorem 2.19. *In the set of C^∞ one-parameter families of diffeomorphisms unfolding a homoclinic tangency at the parameter value $\mu = 0$ there exists a residual set $\mathcal{R}$ such that if $(f_\mu)_\mu$ is in $\mathcal{R}$ then there is a sequence of parameter intervals I_n with $I_n \to 0$ and a dense subset D_n in I_n so that f_μ has infinitely many Hénon-like strange attractors for all $\mu \in D_n$.*

This result says nothing about persistence of the coexistence of these infinitely many attractors. It was an open question whether there exists a k-parameter family $f = (f_a)_a$ which has infinitely many attractor (strange or not) for values of the parameter a in a positive Lebesgue measure set $E(f) \subset \mathbb{R}^k$. Palis claimed that the measure the set $E(f)$ is generically zero for families $f = (f_a)_a$ of one-dimensional dynamics and surface diffeomorphism Palis (2000) and Palis and Takens (1993).

Theorem 2.18 (see also Pumariño and Rodríguez (2001, Thm. B), for diffeomorphisms) provided a first example of a family of dynamical systems X_a with persistent coexistence of infinitely many strange attractor. Since X_a is not generic, this family is not a counterexample to the possible extension of the above conjecture to three-dimensional vector fields. Although the Palis conjecture remains still open, some advances in the opposite direction have been made by Berger in Berger (2016, 2017) for families of surface endomorphisms (in fact, local diffeomorphisms) and higher dimensional diffeomorphisms. Namely, Berger constructed open sets of families of the above described dynamics (in suitable and well-defined topologies) such that residually in these open sets any family exhibits simultaneously infinitely many hyperbolic periodic attractors *for all* parameter value.

The persistence of coexistent strange attractors is physically relevant because it means that these attractors can be seen in applications. If we ignore this property, results of coexistence of infinitely many attractors may be found in the literature. For example, denote by $\mathcal{S}_0$ the codimension one submanifold of dissipative C^∞ vector fields X that have a Shilnikov homoclinic cycle $\Gamma = \Gamma(X)$, i.e., satisfying **(S1)**,**(S2)** and **(S3b)**. In Ovsyannikov and Shilnikov (1987) it was proved the following:

Theorem 2.20. *For each $\varepsilon > 0$, the following sets are dense in $\mathcal{S}_0$:*

(i) *the open set of vector fields X in $\mathcal{S}_0$ that have an attracting 2-periodic hyperbolic orbit in an ε-neighborhood of $\Gamma(X)$;*

(ii) *the set of vector fields X that possess infinitely many attracting 2-periodic orbits in an ε-neighborhood of $\Gamma(X)$;*

(iii) *the set of vector fields X that have an orbit of homoclinic tangency to a hyperbolic periodic orbit in an ε-neighborhood of $\Gamma(X)$.*

Using Colli (1998) and Ovsyannikov and Shilnikov (1987), one can get the following (see Homburg (2002) and Homburg and Sandstede (2010)):

Theorem 2.21. *For each $\varepsilon > 0$ there exists a dense subset $\mathcal{D} \subset \mathcal{S}_0$ such that each vector field $X \in \mathcal{D}$ has infinitely many coexisting strange attractors in a ε-neighborhood of $\Gamma(X)$.*

A stronger result for non-generic unfoldings of Shilnikov homoclinic orbits was obtained by Homburg (2002). For families of vector fields in $\mathcal{S}_0$ Homburg proved the next theorem.

Theorem 2.22. *Consider a one-parameter family X_a of vector fields in $\mathcal{S}_0$ such that*

$$\frac{\partial \delta}{\partial a}(0) \neq 0.$$

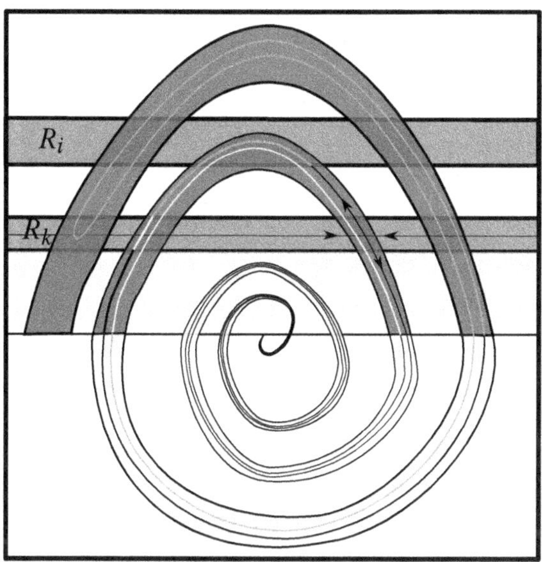

Figure 2.9: *Unstable manifold of a fixed point p of Π_a in R_k.*

Let Σ be a cross-section transverse to the homoclinic cycle Γ_a and Π_a the corresponding first-return map of X_a. Denote by $\mathcal{T}_n$ a decreasing sequence of tubular neighborhoods of Γ_a. Then there is an interval of parameters I containing $a = 0$ such that the following holds for n large enough:

(i) there is an open and dense set P_n of I with Lebesgue measure going to zero as $n \to 0$ for which Π_a has an attracting 2-periodic orbit in $\mathcal{T}_n$. Furthermore, the set of parameter values $a \in I$ for which Π_a has infinitely many 2-periodic attractors is also dense in I but has zero measure.

(ii) there is a dense set E_n of I with positive Lebesgue measure going to zero as $n \to 0$ for which Π_a has a Hénon-like strange attractor in $\mathcal{T}_n$ intersecting Σ in two connected components.

Figure 2.9 shows the unstable manifold of a fixed point p of Π in R_k. The idea of Homburg to prove the above result consisted in observe the interaction of this unstable manifold with other strip R_i so that $\Pi_a(R_k) \cap R_i$ is not fully as illustrated in the same figure. The second iterated of Π_a restricted to R_k maps this strip into a horseshoe shape with the fold inside R_k. This figure shows the possibility of occur homoclinic tangencies and therefore after the unfolding the emergence of attractors by varying the internal parameter δ. However, Homburg does not prove directly here the occurrence of tangencies. He proved that after a renormalization processes around this fold the limit family is a good unfolding of a strong dissipative Hénon map and thus, Hénon-like attractors appear.

On the other hand, Homburg also studied the density of the orbits caught by the attractors. Denote by D_n the set of points $x \in \Sigma \cap \mathcal{T}_n$ in the above theorem for which the positive orbit of x leaves $\mathcal{T}_n$. Then it is proved in Homburg (2002) that

$$\lim_{n \to \infty} \frac{|D_n|}{|\Pi_a \cap U_n|} = 1.$$

This result establishes that the large number of attractors do not trap most orbits in an small enough neighborhood of the Shilnikov cycle Γ. In order to recover the orbits that escape can be assumed the existence of a double Shilnikov loop $\Gamma^{\pm} = \{O\} \cup \gamma^+ \cup \gamma^-$. See Arneodo, Coullet, and Tresser (1981) and also Pacifico, Rovella, and Viana (1998). In the following section we will consider this kind of configurations and we will show that in such a case, the dynamics could be non-trivial even in the simply case of the assumption **(S4)**.

2.6 Switching

As we have seem, the scenario described by the assumptions **(S1)**, **(S2)** and **(S4)** provides *simple dynamics* around the homoclinic cycle $\Gamma = \{O\} \cup \gamma$. However the situation could be different if we assume that (2.1) has a *homoclinic network* (i.e, a finite union of homoclinic orbits associated with the same equilibrium point). Observe that orbits that start in $\Sigma^+ = \Sigma \cap V^+$ will reach Out^+, and then follow the homoclinic orbit γ until they will return once again to Σ. Trajectories that start in $\Sigma^- = \Sigma \cap V^-$ may (or not) return to Σ. As illustrated in Figure 2.10, the existence of a second homoclinic orbit allows to introduce another different global map and thus we will have two different first-return maps to Σ. Therefore, we will assume the following hypothesis.

(S2') There are two different trajectories γ^+ and γ^- biasymptotic to
O.

The homoclinic network is given by $\Gamma^{\pm} = \{O\} \cup \gamma^+ \cup \gamma^-$. This kind of configuration usually holds under symmetry assumption. For instance, if system (2.1) satisfies **(S2)** and f is $\mathbb{Z}_2$-symmetric under the action of $-\mathrm{Id}$. That is,

$$f(-x) = -f(x) \quad \text{for all } x \in \mathbb{R}^3.$$

In this case, the additional homoclinic orbit is given by $\gamma^- = -\mathrm{Id}(\gamma^+)$ where $\gamma^+ = \gamma$. We address the reader to Golubitsky, Stewart, and Schaeffer (2012) for more details about equations with symmetry.

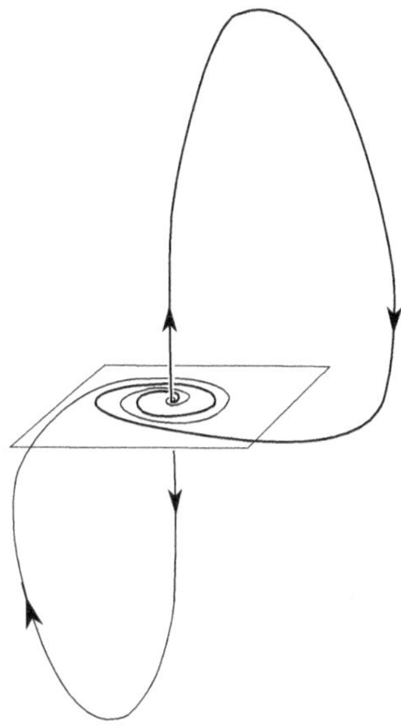

Figure 2.10: *If the orbit returns to Σ^-, then it is reinjected following another different homoclinic orbit.*

The goal of this section is to show that under the assumptions **(S1)**, **(S2')** and **(S4)**, the dynamics that remain forward is a tubular neighborhood of the network $\Gamma^{\pm}$ still could be very complicated. To do this we introduce the notion of switching in a general context. That is, for any homoclinic network

$$\Gamma = \{O\} \cup \gamma_1 \cup \cdots \cup \gamma_m$$

of a differential equation associated with a equilibrium point O with $m \geq 2$. Associated with this network, we consider small cross-sections S_i transverse to γ_i. Let us denote by Π the first-return map on the collection of these pairwise disjoint cross-sections

$$S = S_1 \cup \cdots \cup S_m$$

defined by following the homoclinic orbit γ_i if the initial condition is in S_i.

Definition 2.23. *The homoclinic network Γ exhibits* infinite switching *if for each tubular neighborhood $\mathcal{T}$ of Γ and every $\omega = (\omega_i)_{i \in \mathbb{N}} \in \{1, \ldots, m\}^{\mathbb{N}}$, there exists a flow trajectory τ and a point $x_\omega = \tau(t_0)$ for some time t_0 such that*

$$\tau(t) \in \mathcal{T} \text{ for all } t \geq t_0 \quad \text{and} \quad \Pi^i(x_\omega) \in S_{\omega_{i+1}} \text{ for all } i \geq 0.$$

We call x_ω the starting point of the realization τ for the infinite path $\omega^\infty = (\gamma_{\omega_i})_{i \in \mathbb{N}}$ of homoclinic connections.

An infinite path on Γ can be considered as a pseudo-orbit of (2.1) with infinitely many discontinuities. Switching near Γ means that any pseudo-orbit in Γ can be realized. For another equivalent definition and extension of the notion of switching see Aguiar, Labouriau, and Rodrigues (2010), Homburg and Knobloch (2010), Ibáñez and Rodrigues (2015), and Rodrigues (2016). Observe that if Γ exhibit infinite switching (or switching for short) then in any tubular neighborhood $\mathcal{T}$ of the network we find a set of forward orbits so that Π restrict to the set of starting points of this trajectories semi-conjugate to the full unilateral shift on m symbols. Because of this, the first-return map restricted to this set of orbits has positive topologically entropy which can be seen as an indicator of *chaos*.

Complex behavior near a homoclinic network is often connected to the occurrence of switching, a term which has also been used to describe simpler dynamics where there is one change in the choices observed in trajectories. This is the case described in Kirk and Silber (1994). In fact, there are several examples in the literature where the existence of infinite switching leads to complicate behavior near the network, see Aguiar, Labouriau, and Rodrigues (2010), Labouriau and Rodrigues (2012), Rodrigues (2013a), and Rodrigues and Labouriau (2014). In general, infinite switching is related with the existence of suspended horseshoes in its neighborhood. See for example the works Ibáñez and Rodrigues (2015), Labouriau and Rodrigues (2012), and Rodrigues (2013b). In these articles, the authors proved the existence of infinitely many initial conditions that realize a given forward infinite path. These solutions lie on the sequence of suspended horseshoes that accumulate on the network. So, one could think a priori that infinite switching seems to be connected with the existence of suspended horseshoes. However in the following result we will see that this is not necessarily true.

Theorem 2.24 (Holmes (1980), Rodrigues (2016), and Tresser (1984)). *Under the assumptions (S1), (S2') and (S4) the following conditions hold:*

   (i) there are no suspended horseshoes in any small neighborhood of the homoclinic network $\Gamma^{\pm}$;

   (ii) the homoclinic network $\Gamma^{\pm}$ attracts all the trajectories in a small neighborhood of the network;

   (iii) there is infinite switching near the homoclinic network $\Gamma^{\pm}$.

Indeed, item (i) is a immediately consequence of item (ii). On the hand, item (ii) follows from Theorem 2.6 since now the trajectories starting near the network remain close to $\Gamma^{\pm}$ in forward time. In what follows we will prove item (iii) following Rodrigues (2016).

First of all, similarly to §2.2, we consider a cross-section Σ in V transversal to both γ^+ and γ^- at points q^+ and q^- respectively. We write $\Sigma^+ = \Sigma \cap V^+$ and $\Sigma^- = \Sigma \cap V^-$. For each homoclinic orbit γ^+ and γ^- we can construct a first-return map Π to Σ starting on Σ^+ and Σ^- and following the corresponding trajectory respectively. This first-return consists in two maps that could be modeled as in (2.8). The difference between these two maps is only in the global map. However, as before, still the global map from Out^- to Σ preserves the spiraling shape on Out^- which is obtained by the method described in Lemma 2.3. This property will be the unique tool that we need.

Lemma 2.25. *For any vertical line β on Σ there are sequences $(\beta_i^{\pm})_{i \in \mathbb{N}}$ of non-empty compact subsets of β accumulating on $\beta \cap W_{loc}^s(O)$ such that $\Pi(\beta_i^{\pm})$ is a vertical line in $\Sigma^{\pm}$ for all $i \in \mathbb{N}$.*

Proof. By Lemma 2.3, the image of β under Φ is a spiral accumulating on $\mathrm{Out} \cap W_{loc}^u(O)$. In its turn, this spiral is mapped by the global map into another spiral in Σ accumulating on $\Gamma^{\pm} \cap \Sigma$. The curve $W_{loc}^s(O)$ cuts transversely this spiral into infinitely many points. Thus we find infinitely many vertical lines $\widetilde{\beta}_i^{\pm}$ on $\Sigma^{\pm}$ contained in this spiral as illustratively is showed in Figure 2.11. Consider the non-empty compact set $\beta_i^{\pm} = \Pi^{-1}(\widetilde{\beta}_i^{\pm}) \subset \beta$. By construction, it follows that $\Pi(\beta_i^{\pm})$ is a vertical line on $\Sigma^{\pm}$. $\qquad\square$

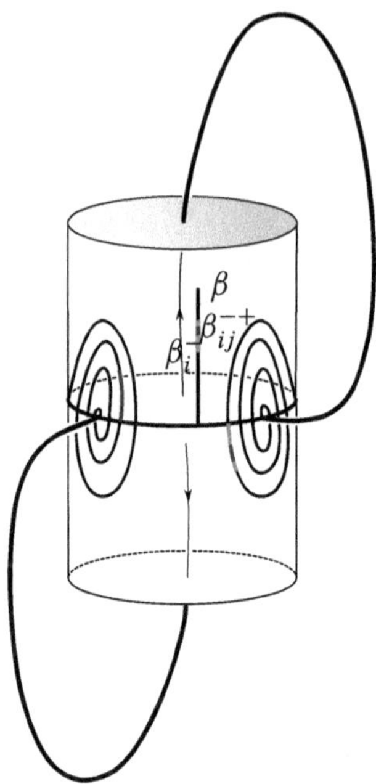

Figure 2.11: *Illustration of the inductive construction $\beta_{i_1\dots i_n}^{j_1\dots j_n}$.*

By an inductive process using the above lemma, we get sequences of non-empty compact sets $\beta_{i_1\dots i_n}^{j_1\dots j_n}$ where $i_1,\dots,i_n \in \mathbb{N}$ and $j_1,\dots,j_n \in \{+,-\}$ for all $n \in \mathbb{N}$ such that

- $\beta_{i_1\dots i_n}^{j_1\dots j_n} \subset \beta_{i_1\dots i_{n-1}}^{j_1\dots j_{n-1}} \subset \beta,$

- $\Pi^n(\beta_{i_1\dots i_n}^{j_1\dots j_n})$ is a vertical line on Σ^{j_n}.

Indeed, the case $n = 1$ follows immediately from Lemma 2.25. Assume constructed $\beta_{i_1\dots i_n}^{j_1\dots j_n}$ for $n \geqslant 1$. Hence, since $\Pi^n(\beta_{i_1\dots i_n}^{j_1\dots j_n})$ is a vertical line on Σ, applying again Lemma 2.25 one finds sequences of non-empty compact sets

$$\widetilde{\beta}_{i_1\dots i_n i_{n+1}}^{j_1\dots j_n j_{n+1}} \subset \Pi^n(\beta_{i_1\dots i_n}^{j_1\dots j_n}) \quad \text{for } i_{n+1} \in \mathbb{N} \text{ and } j_{n+1} \in \{+,-\}$$

which are sent by the first-return map into a vertical line on $\Sigma^{j_{n+1}}$. Therefore, taking

$$\beta_{i_1\dots i_n i_{n+1}}^{j_1\dots j_n j_{n+1}} = \Pi^{-n}(\widetilde{\beta}_{i_1\dots i_n i_{n+1}}^{j_1\dots j_n j_{n+1}})$$

it is easy to see that the required properties hold.

Finally, we will prove item (iii) in Theorem 2.24. Fix any small tubular neighborhood $\mathcal{T}$ of $\Gamma^{\pm}$. Consider the section $S^{\pm} = \mathcal{T} \cap \Sigma^{\pm}$. We restrict the above return map Π to $S = S^{+} \cup S^{-}$. By the attracting character of the network, trajectories starting in S remain in $\mathcal{T}$ for all positive time. Let us take sequences $\omega = (\omega_n)_{n \in \mathbb{N}} \in \{+, -\}^{\mathbb{N}}$ and $i = (i_n)_{n \in \mathbb{N}}$ where $i_n \in \mathbb{N}$. Consider $j = (j_n)_{n \in \mathbb{N}}$ with $j_n = \omega_{n+1}$ for all $n \in \mathbb{N}$ and a vertical line β on S^{ω_1}. By the inductive construction above, we get a non-empty compact set

$$\beta_i^j = \bigcap_{n=1}^{\infty} \beta_{i_1 \ldots i_n}^{j_1 \ldots j_n} \subset \beta$$

such that for every $x \in \beta_i^j$ holds that

$$x \in \beta \subset S^{\omega_1} \quad \text{and} \quad \Pi^n(x) \in \Pi^n(\beta_{i_1 \ldots i_n}^{j_1 \ldots j_n}) \subset S^{j_n} = S^{\omega_{n+1}} \quad \text{for all } n \in \mathbb{N}.$$

This proves that $\Gamma^{\pm}$ exhibits infinite switching.

3   *Bykov cycles*

We have seen in the previous chapter that the existence of Shilnikov homoclinic orbits provides one of the main criteria for the occurrence of chaos involving suspended horseshoes and non-hyperbolic strange attractors. In this section, we study Bykov cycles in a three-dimensional phase space, which consist of heteroclinic cycles associated to two saddle-foci with different Morse indices (dimension of the unstable manifold); one of the heteroclinic connections is determined by the intersection of two-dimensional invariant manifolds and it is assumed to be transverse. The other heteroclinic connection corresponds to the coincidence of two branches of the one-dimensional invariant manifolds. These cycles were first studied by Bykov (1999, 2000) (see also Fernández-Sánchez, Freire, Pizarro, et al. (2003), Fernández-Sánchez, Freire, and Rodríguez-Luis (2002), and Glendinning and Sparrow (1986)) and recently there has been a renewal of interest on this type of heteroclinic bifurcation in different contexts. Other types of cycles associated to equilibria whose linearization of the vector field has real eigenvalues were previously explored in Bykov (1993) and Kokubu (1991). These cycles are associated with intermittent dynamics and are used to model stop-and-go behaviour in various applications.

As shown in Bykov (1999, 2000), spirals in a two-dimensional parameter space are organized around T[erminal]-points corresponding to a codimension two Bykov cycle. For this reason, Bykov cycles are also known as T-points. The spirals are related with Shilnikov homoclinic bifurcations and have been located in several models from many fields. Bykov cycles arise in the Lorenz equations Glendinning and Sparrow (1986), (reversible) systems that model the propagation of calcium waves, Michelson system Dumortier, Ibáñez, and Kokubu (2001a), Ibáñez and Rodríguez (2005), and Lamb, Teixeira, and Webster (2005), Josephson junctions Van den Berg, van Gils, and Visser (2003), electronic oscillators Fernández-Sánchez, Freire, and Rodríguez-Luis (2002), Hopf–zero bifurcation Dumortier, Ibánez, Kokubu, and Simó (2013) and Lamb, Teixeira, and Webster (2005) and in generic unfoldings of nilpotent singularities of codimension three Barrientos, Ibáñez, and Rodríguez (2011) and Ibáñez and Rodríguez (2005). See also the approaches of Labouriau and Rodrigues (2015), Lamb, Teixeira, and Webster (2005), Oldeman, Krauskopf, and Champneys (2001), and Rodrigues (2013b) and references therein.

The comparison between the two ways the flow turns around the two saddle-foci determines two types of Bykov cycles: either both saddle-foci have the same *chirality* or it is different. We will study the recurrent dynamics in both cases. The existence of horseshoes for the return map follows under both assumptions. However, when chirality is different, homoclinic tangencies between the invariant manifolds of hyperbolic fixed points are also exhibited. This means that, when the chirality is different, persistent strange attractors do exist in a neighborhood of the cycle. Existence of Shilnikov homoclinic orbits in generic unfoldings will be also argued.

3.1　Setting

Consider a differential equation

$$\dot{u} = X(u), \qquad u \in \mathbb{R}^3 \tag{3.1}$$

where X is a C^r vector field with $r > 1$. We assume that the associated flow satisfies the following hypotheses:

(BC1) There are two different equilibria O_1 and O_2 with eigenvalues of $DX(O_1)$ and $DX(O_2)$ respectively,

$$\lambda_1, \ -\alpha_1 \pm \omega_1 i \qquad \text{and} \qquad -\lambda_2, \ \alpha_2 \pm \omega_2 i$$

where $\lambda_j, \alpha_j > 0$ for $j = 1, 2$ and $\omega_1 \omega_2 \neq 0$.

(BC2) There are two heteroclinic orbits $\gamma_{1 \to 2}$ and $\gamma_{2 \to 1}$ such that

$$\gamma_{1 \to 2} \subset W^u(O_1) \cap W^s(O_2) \quad \text{and} \quad \gamma_{2 \to 1} \subset W^u(O_2) \pitchfork W^s(O_1),$$

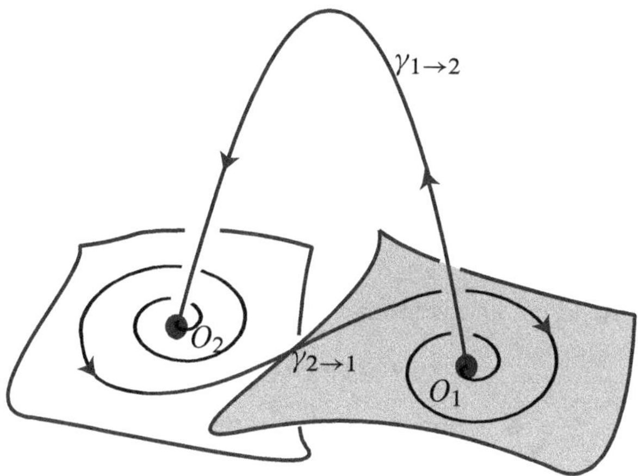

Figure 3.1: *Illustration of a Bykov cycle.*

where the symbol $\pitchfork$ means that the two manifolds $W^u(O_1)$ and $W^s(O_2)$ meet transversely.

The assumption **(BC2)** asks that the one-dimensional manifold $W^u(O_1)$ and $W^s(O_2)$ meets along $\gamma_{1\to 2}$ while the two-dimensional manifolds $W^u(O_2)$ and $W^s(O_1)$ intersect transversely along one trajectory $\gamma_{2\to 1}$. The heteroclinic cycle

$$\Gamma = \{O_1, O_2\} \cup \gamma_{1\to 2} \cup \gamma_{2\to 1}$$

is illustrated in Figure 3.1 and we refer to it as a *Bykov cycle*.

3.2 Return maps

We will analyze the dynamics near Γ, using local maps that approximate the dynamics near and between the two equilibria in the cycle.

Cross-sections. We need a workable expression of the Poincaré map at a suitable cross section inside a tubular neighborhood of the Bykov cycle. First of all, as in §2.2, since the vector field is regular enough, we can consider linear C^1 coordinates around O_1 and O_2. In these coordinates, we consider two cylindrical neighborhoods V_1 and V_2 of O_1 and O_2, respectively, with base-radius ε and height $2\varepsilon > 0$ (see Figure 3.2). Their boundaries consist of three components:

- The cylinder wall parametrized by $x \in \mathbb{S}^1$ and $|y| \leqslant \varepsilon$ with the usual cover

$$(x, y) \mapsto (\varepsilon, x, y).$$

Here y represents the height of the cylinder and x is the angular coordinate, measured from the point $x = 0$ in the intersection of the connection $\gamma_{2 \to 1}$ with the wall of the cylinder. In these coordinates the local stable and local unstable manifolds of O_1 and O_2 correspond to $y = 0$, respectively.

- Two discs, the top and the bottom of the cylinder. We assume that the connection $\gamma_{1 \to 2}$ goes from the top of one cylinder V_1 to the top of the other, and we take a polar covering of the top disc

$$(r, \varphi) \mapsto (r, \varphi, \varepsilon)$$

where $0 \leqslant r \leqslant \varepsilon$ and $\varphi \in \mathbb{S}^1$. In these coordinates the local unstable and local stable manifolds of O_1 and O_2, respectively, correspond to $r = 0$.

Consider the cylinder wall of V_1. As mentioned, the local stable manifold $W^s_{loc}(O_1)$ meets this wall at the circle parametrized by $y = 0$. The top part of the wall of V_1 is denoted by $\text{In}^+(O_1)$ and corresponds to points with $y > 0$. Trajectories starting at interior points of $\text{In}^+(O_1)$ go into the cylinder in positive time and come out through the top disc, denoted $\text{Out}^+(O_1)$. Trajectories starting at interior points of $\text{Out}^+(O_1)$ go inside the cylinder in negative time. Moreover, $W^u_{loc}(O_1) \cap \text{Out}^+(O_1)$ corresponds with the positive z-axis intersecting $\text{Out}^+(O_1)$ at the origin of coordinates $r = 0$. Analogous results hold for $y < 0$.

Reversing the time, we get dual results for O_2. In this case, the z-axis correspond to $W^s_{loc}(O_2)$ and meets the top disc of the cylinder, $\text{In}^+(O_2)$, at the origin of its coordinates. Moreover, the intersection of $W^u_{loc}(O_2)$ with V_2 is parametrized by the circle $y = 0$. Trajectories starting at interior points of $\text{In}^+(O_2)$ go into V_2 in positive time and leave the cylinder through its wall with $y \geqslant 0$, that we denote by $\text{Out}^+(O_2)$. Analogous results hold for $y < 0$.

Local maps. In cylindrical coordinates (ρ, θ, z), the linearization of the dynamics at O_1 and O_2 is specified by the following equations:

$$\begin{cases} \dot{\rho} = -\alpha_1 \rho \\ \dot{\theta} = \omega_1 \\ \dot{z} = \lambda_1 z \end{cases} \quad \text{and} \quad \begin{cases} \dot{\rho} = \alpha_2 \rho \\ \dot{\theta} = \omega_2 \\ \dot{z} = -\lambda_2 z. \end{cases} \qquad (3.2)$$

The above systems are like those studied in (see § 2.2). Rescaling the variables in order to take $\varepsilon = 1$, it easily follows that a trajectory whose initial point is $(x, y) \in \text{In}(O_1) \setminus W^s_{loc}(O_1)$ leaves V_1 through $\text{Out}(O_1)$ at the point

$$(r, \varphi) = \Phi_1(x, y) = \left(y^{\delta_1}, \ -\frac{\omega_1}{\lambda_1} \log y + x \right) \qquad (3.3)$$

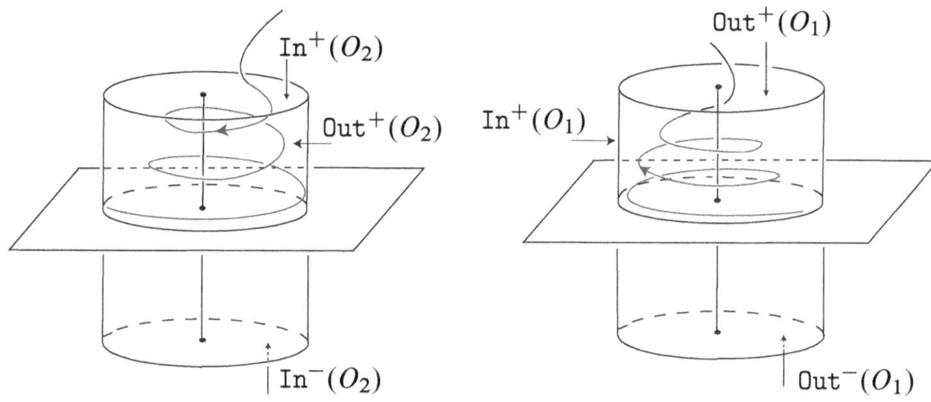

Figure 3.2: *Local cylindrical coordinates in V_1 and V_2, near O_1 and O_2.*

where $\delta_1 = \alpha_1/\lambda_1 > 0$. By reversing the time, the analysis of the second system in (3.2) is analogous and thus a point $(r, \varphi) \in \text{In}(O_2) \setminus W_{loc}^u(O_2)$ leaves V_2 through $\text{Out}(O_2)$ at the point

$$(x, y) = \Phi_2(r, \varphi) = \left(-\frac{\omega_2}{\alpha_2} \log r + \varphi, \; r^{\delta_2} \right) \tag{3.4}$$

with $\delta_2 = \lambda_2/\alpha_2 > 0$.

Global maps. As shown in Figure 3.3, points in $\text{Out}^+(O_1)$ near $W^u(O_1)$ are mapped into $\text{In}^+(O_2)$ along a flow-box around the connection $\gamma_{1\to2}$. Up to a change of coordinates, corresponding to homotheties and rotations which leave invariant the local expressions of the flows in neighborhoods of the equilibria, consistently with hypothesis **(BC2)**, we may assume that the transition

$$\Psi_{1\to2} : \text{Out}^+(O_1) \to \text{In}^+(O_2)$$

is a linear map given, in rectangular coordinates $X = r \cos \varphi$ and $Y = r \sin \varphi$, by:

$$(\bar{X}, \bar{Y}) = \Psi_{1\to2}(X, Y) = (aX, a^{-1}Y) \tag{3.5}$$

for some constant $a \geqslant 1$. Note that the map $\Psi_{1\to2}$ is given in rectangular coordinates. To compose this map with Φ_2, it is required to change again the coordinates. Indeed, in appropriate coordinates $(\bar{r}, \bar{\varphi}) = \Psi_{2\to1}(r, \varphi)$, we get:

$$\bar{r} = r\sqrt{a^2 \cos^2 \varphi + a^{-2} \sin^2 \varphi} \quad \text{and} \quad \bar{\varphi} = \arctan\left(a^2 \tan \varphi\right).$$

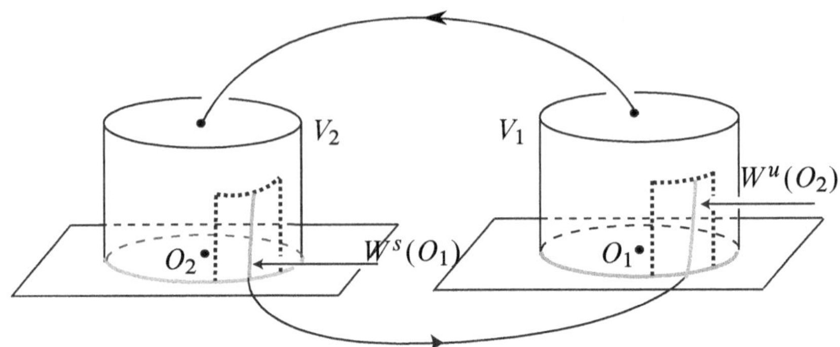

Figure 3.3: *Transition map.*

Note that a circle of radius $r < 1$ in $\mathtt{Out}^+(O_1)$, centered at the origin, is mapped by $\Psi_{1\to 2}$ into an ellipse centered at the origin of $\mathtt{In}^+(O_2)$ with major axis of length $a\,r \geqslant r$ and minor axis of length $r/a \leqslant r$.

On the other hand, as we may observe in Figure 3.3, in $\mathtt{In}(O_1)$, the set $W^u(O_2)$ is now described by a curve of the form $\beta(s) = (0, s)$. In particular, the set $W^u(O_2)$ crosses $\mathtt{In}(O_1)$ from the top to the bottom. In view of this, we may suppose that the map

$$\Psi_{2\to 1} : \mathtt{Out}(O_2) \longrightarrow \mathtt{In}(O_1)$$

is given, in the suitable coordinates of $\mathtt{Out}(O_2)$ and $\mathtt{In}(O_1)$, by

$$(\bar{x}, \bar{y}) = \Psi_{2\to 1}(x, y) = (y, -x). \tag{3.6}$$

The Poincaré map   Let

$$\Pi = \Psi_{2\to 1} \circ \eta : \mathtt{In}(O_1)\backslash W^s_{\mathrm{loc}}(O_1) \to \mathtt{In}(O_1) \tag{3.7}$$

with

$$\eta = \Phi_2 \circ \Psi_{1\to 2} \circ \Phi_1 : \mathtt{In}(O_1)\backslash W^s_{\mathrm{loc}}(O_1) \to \mathtt{Out}(O_2),$$

be the first return map to $\mathtt{In}(O_1)$. For $a = 1$, by a straightforward calculation, it is not difficult to see that the map Π may be written as

$$\Pi(x, y) = \left(y^\delta, \ K_\omega \ln y - x \right). \tag{3.8}$$

where

$$K_\omega = \frac{\omega_2\,\alpha_1 + \omega_1\,\alpha_2}{\lambda_1\,\alpha_2} \quad \text{and} \quad \delta = \delta_1\delta_2. \tag{3.9}$$

(a) Different chirality.　　　　　　　(b) Same chirality.

Figure 3.4: *Illustration of a Bykov cycle with the two kinds of chirality.*

Chirality   There are two different possibilities for the geometry of the flow around Γ, depending on the direction trajectories turn around the heteroclinic connection from O_1 to O_2.

Let V_1 and V_2 be small disjoint neighbourhoods of O_1 and O_2 with disjoint boundaries ∂V_1 and ∂V_2, respectively. Trajectories starting at ∂V_1 near $W^s(O_1)$ go into the interior of V_1 in positive time, then follow the connection from O_1 to O_2, go inside V_2, and then come out at ∂V_2 as in Figure 3.4. Let Φ be a piece of trajectory like this from ∂V_1 to ∂V_2. Now join its starting point to its end point by a line segment as in Figure 3.4, forming a closed curve, that we call *loop*. Observe that this loop and the cycle Γ are disjoint closed sets.

Definition 3.1. *We say that the two saddle-foci O_1 and O_2 in Γ have the* same chirality *if the loop of every trajectory is linked to Γ in the sense that the two closed sets cannot be disconnected by an isotopy. Otherwise, we say that O_1 and O_2 have* different chirality.

We will study Bykov cycles under the following two assumptions on the chirality (see Figure 3.4):

(BC3)   The saddle-foci O_1 and O_2 have the same chirality.

($\overline{\textbf{BC3}}$)   The saddle-foci O_1 and O_2 have different chirality.

Under hypotheses **(BC1)**, **(BC2)** and **(BC3)**, we will show that, in a small neighborhood of the Bykov cycle, there exists a sequence of uniformly hyperbolic suspended horseshoes accumulating on the cycle. The periodic solutions giving n loops around the cycle will accumulate on special trajectories called n-pulses (compare with Definition 2.10). "Far" from the cycle, under additional hypotheses on the way the cycle is obtained, the authors of Labouriau and Rodrigues (2016), have shown the existence of homoclinic tangencies to a periodic solution and strange attractors.

On the other hand, under hypotheses **(BC1)**, **(BC2)** and $(\overline{\textbf{BC3}})$, we will show that, in a small neighborhood of the Bykov cycle, besides the sequence of uniformly hyperbolic suspended horseshoes and n-pulses which accumulate on the cycle, we also find heteroclinic tangencies, sinks and strange attractors. Different chirality of the equilibria is quite involving and brings richer dynamics to the problem.

3.3   Bykov cycles with same chirality

In the present section, we consider a class of vector fields satisfying hypotheses **(BC1)**–**(BC3)**. That is, we study Bykov cycles where the saddle-foci have the same chirality. These cycles appear in the Michelson (reversible) system Dumortier, Ibánez, and Kokubu (2005), Ibáñez and Rodríguez (2005), and Lamb, Teixeira, and Webster (2005), in the reversible Hopf-zero bifurcation Dumortier, Ibánez, Kokubu, and Simó (2013), in $\mathbb{Z}^2$ symmetric systems Fernández-Sánchez, Freire, and Rodríguez-Luis (2002) and in generic unfoldings of any three-dimensional nilpotent singularity of codimension three Barrientos, Ibáñez, and Rodríguez (2011) and Ibáñez and Rodríguez (2005). See also Chapter 5.

In this section, assuming that $a = 1$ is not a loss of generality Rodrigues (2013b). The dynamics near a Bykov cycle where the saddle-foci have the same chirality is characterized by the existence of infinitely many suspended horseshoes accumulating on the cycle. Another characteristic is the presence of infinitely many heteroclinic orbits from O_2 to O_1. Namely, for each $n \in \mathbb{N}$ we will find a heteroclinic orbit hitting a cross-section to the cycle at precisely $n \in \mathbb{N}$ points. These trajectories are called n-pulse heteroclinic orbits.

Theorem 3.2 (Bykov (2000), Knobloch, Lamb, and Webster (2014), and Labouriau and Rodrigues (2012)). *Under the assumptions **(BC1)**,**(BC2)** and **(BC3)**, any tubular neighborhood $\mathcal{T}$ of the Bykov cycle Γ contains the following:*

1. *infinitely many n-pulses heteroclinic orbits from O_2 to O_1, for each $n \in \mathbb{N}$;*

2. *a sequence of hyperbolic compact Π-invariant sets Λ_k in $\mathrm{In}(O_1)$ such that the restriction of Π to Λ_k is conjugate to the full shift of k symbols, for $k \geqslant 2$. Moreover, Π restricted to the union Λ of the sets Λ_k is topologically conjugate to a full shift over an infinite number of symbols.*

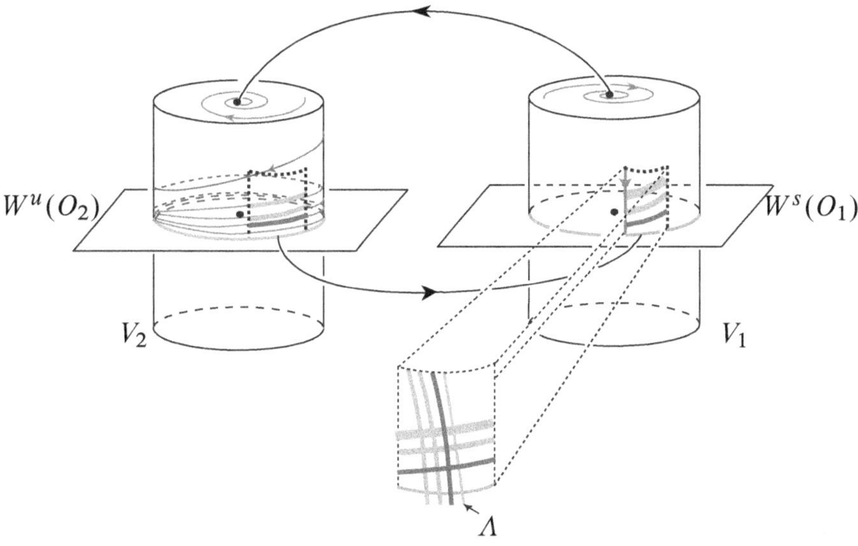

Figure 3.5: *Chain of horseshoes near the Bykov cycle.*

We sketch the proof of Theorem 3.2. Item (1) is a direct consequence of Aguiar, Labouriau, and Rodrigues (2010), picking the segment β as $W^u(O_2) \cap \text{In}(O_1)$. The image of the segment β, under the Poincaré map, contains infinitely many segments accumulating on itself and crossing transversally $W^s(O_1) \cap \text{In}(O_1)$.

Now we prove item (2). If $R_1 \subset \text{In}(O_1)$ is a rectangle containing $\gamma_{2\to1}$ on its border as sketched in Figure 3.5, the initial conditions that returns to $\text{In}(O_1)$ are contained in a sequence of horizontal strips accumulating on the stable manifold of O_1, whose heights tend to zero Rodrigues (2013b). Each one of these horizontal strips lying on the rectangle $R_1 \subset \text{In}(O_1)$, is mapped under $\eta = \Phi_2 \circ \Psi_{1\to2} \circ \Phi_1$ into a horizontal strip across $\text{Out}(O_2)$. By **(BC2)**, they are mapped by $\Psi_{2\to1}$ into vertical strips across R_1 crossing transversely the original rectangles. This gives rise to a sequence of uniformly hyperbolic horseshoes, accumulating on the heteroclinic cycle Γ.

Although the results of Lamb, Teixeira, and Webster (2005) used reversibility, many of the results on the dynamics near the heteroclinic cycle do not rely on the reversibility of the system. Their results on the existence of horseshoes, similar to our Proposition 3.2, are independent of any Shilnikov condition on the saddle-values since the proof relies only on the geometry of the flow near the cycle.

The countable family of horseshoes $(\Lambda_k)_{k\in\mathbb{N}}$, whose union builds the set

Λ, is contained in a compact set inside $\text{In}(O_1)$. Therefore, given $k \in \mathbb{N}$, any sequence p_n^k in Λ of periodic points with period k by the Poincaré map Π has accumulation points. Next result clarifies these accumulation points:

Theorem 3.3 (Bessa, Carvalho, and Rodrigues (2017)). *Any accumulation point of p_n^k lies in $W^s(O_1) \cap W^u(O_2)$ and its solution by the flow associated with X is a $(k-1)$–pulse connection.*

In Bessa, Carvalho, and Rodrigues (2017) and Labouriau and Rodrigues (2016), the authors extend the description of the dynamics around a Bykov cycle where the saddle-foci have the same chirality, but in the context of a network consisting of two Bykov cycles.

3.4   Bykov cycles with different chirality

In this section, we consider the set $\mathfrak{B}$ of C^r vector fields, with $r > 1$, satisfying the hypotheses **(BC1)**, **(BC2)** and $\overline{\textbf{(BC3)}}$. That is, our object of study is the set of Bykov cycle whose saddle-foci have different chirality.

Remark 3.4. *As we show in §5.6, Bykov cycles where the saddle-foci have the same chirality are exhibited in generic unfoldings of three-dimensional nilpotent singularities of codimension three. As far as we know, no singularity has been discovered unfolding Bykov cycles where the saddle-foci have different chirality.*

In this case, Theorem 3.2 is still valid, but the proof is not so straightforward as Theorem 3.2 because the horizontal border of the strips may reverse the orientation. The complete and detailed proof is performed in §5.3 of Labouriau and Rodrigues (2015).

The next result says that heteroclinic tangencies occur densely, when the parameters that determine the linear part of the vector field at the equilibria lie in a set of full Lebesgue measure (see (3.11)). The tangencies lie near the cycle, in contrast to the findings of Knobloch, Lamb, and Webster (2005) and Lamb, Teixeira, and Webster (2005) for cycles with the same chirality.

Theorem 3.5 (Labouriau and Rodrigues (2015)). *There is an open set $\mathfrak{U}$ of vector fields within $\mathfrak{B}$ such that for any $X \in \mathfrak{U}$, and any tubular neighborhood $\mathcal{T}$ of the Bykov cycle, there exists $Y \in \mathfrak{B}$ arbitrarily C^r close to Y for which $W^u(O_2)$ and $W^s(O_1)$ have a tangency inside $\mathcal{T}$.*

We sketch the main steps of the proof of Theorem 3.5. It relies on the geometry of the map

$$\eta = \Phi_2 \circ \Psi_{1 \to 2} \circ \Phi_1 : \text{In}(O_1) \to \text{Out}(O_2).$$

Denote by $\beta(s) = (0, s) \in \mathrm{In}(O_1)$, $s > 0$, a vertical segment in $\mathrm{In}(O_1)$. Let

$$(x_\beta(s), y_\beta(s)) = \eta(\beta(s)) = \Phi_2 \circ \Psi_{1\to 2} \circ \Phi_1(\beta(s)).$$

Taking into account (3.3), (3.5) and (3.4), we get that

$$x_\beta(s) = -g_2\delta_1 \ln s - \frac{g_2}{2} \ln C(\bar{\varphi}) + \bar{\varphi} \quad \text{and} \quad y_\beta(s) = s^\delta C(\varphi)^{\frac{\delta_2}{2}} \quad (3.10)$$

where

$$\delta_1 = \frac{\alpha_1}{\lambda_1} \qquad \delta_2 = \frac{\lambda_2}{\alpha_2} \qquad \delta = \delta_1\delta_2 \qquad g_1 = \frac{\omega_1}{\mu_1} \qquad g_2 = \frac{\omega_2}{\alpha_2}$$

and

$$\varphi = -g_1 \ln s, \quad C(\varphi) = a^2 \cos^2(\varphi) + a^{-2} \sin^2(\varphi) \quad \bar{\varphi} = \arctan\left(a^2 \tan(\varphi)\right).$$

Remark 3.6. *The hypothesis ($\overline{BC3}$) may be introduced, in local linear equations (3.2), assuming that $\omega_1 > 0$ and $\omega_2 < 0$. Thus, $g_1 > 0$ and $g_2 < 0$.*

Lemma 3.7 (Labouriau and Rodrigues (2015)). *Let $b = \frac{|\omega_2|}{\omega_1}\frac{\alpha_1}{\alpha_2}$. The following assertions hold:*

1. *if $a = 1$, then $x_\beta(s)$ and $y_\beta(s)$ are both monotonic functions of s;*

2. *if $a > 1$, then $y_\beta(s)$ is not a monotonic function of s;*

3. *$\lim_{s\to 0+} y_\beta(s) = 0$ for all $a \geqslant 1$;*

4. *if $b > 1$, then $\lim_{s\to 0+} x_\beta(s) = -\infty$;*

5. *if $0 < b < 1$, then $\lim_{s\to 0+} x_\beta(s) = \infty$.*

The authors of Labouriau and Rodrigues (2015) noticed that, when $a > 1$, the coordinate map x_β is not a monotonic function of s because the curve $\eta \circ \beta$ reverses the direction of its turning around $\mathrm{Out}(O_2)$. This reversion happens infinitely many times. This is the notion illustrated in Figure 3.6 and formalized in the following definition:

Definition 3.8. *We say that a vector field X in $\mathfrak{B}$ has the* dense reversals property *if for the vertical segment $\beta(s) = (0, s) \in \mathrm{In}(O_1)$, $s > 0$, the projection into $W_{loc}^u(O_2)$ of the points where $\eta \circ \beta$ has a vertical tangency, is dense in $W_{loc}^u(O_2) \cap \mathrm{Out}(O_2)$.*

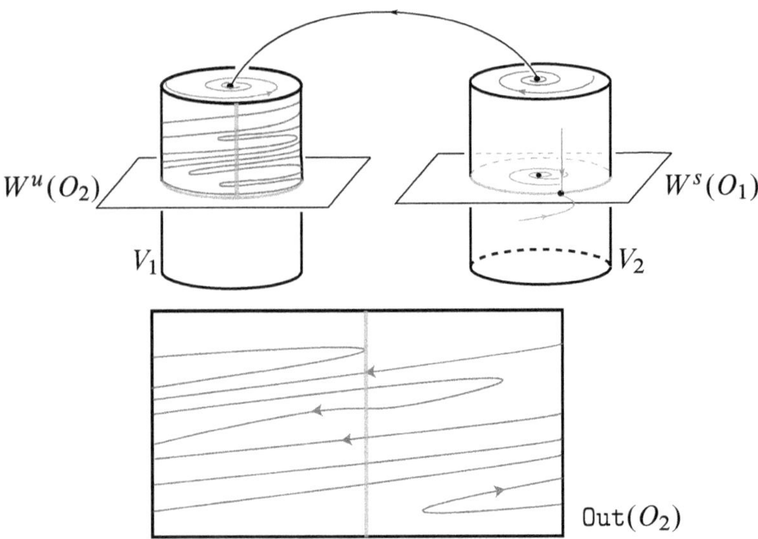

Figure 3.6: *The birth of tangencies for cycles with different chirality.*

Note that Case 1 of Lemma 3.7 rules out reversals when $a = 1$. We need some additional assumptions on the intrinsic parameters of a Bykov cycle

$$P = (\omega_1, \alpha_1, \mu_1, |\omega_2|, \mu_2, \alpha_2, a),$$

determined by the eigenvalues of the linearization of the vector field at the equilibria and the constant $a \geq 1$ which appears in (3.5). The open set $\mathcal{U}$ in Theorem 3.5 corresponds to an open set $\mathfrak{C}$ of parameters P Namely,

$$\mathfrak{C} = \left\{ P : \frac{2\omega_1(a^2 - a^{-2})}{\alpha_1 - \sqrt{\omega_1^2 + 4\alpha_1^2}} < \frac{\alpha_2}{|\omega_2|} - \frac{a^2\alpha_1}{\omega_1} < \frac{2\omega_1(a^2 - a^{-2})}{\alpha_1 + \sqrt{\omega_1^2 + 4\alpha_1^2}} \right\}.$$

$$(3.11)$$

Now define the set $\mathfrak{D}$ as $\{P \in \mathfrak{C} : b \notin \mathbb{Q}\}$, where b was given in the statement of Lemma 3.7. The next result, which concludes the proof of Theorem 3.5, shows that the dense reversals property is a *persistent property* in $\mathfrak{B}$.

Lemma 3.9 (Labouriau and Rodrigues (2015)). *Let $X \in \mathfrak{B}$ with intrinsic parameters $P \in \mathfrak{D}$, then X has the* dense reversals property.

The same idea to prove Theorem 3.5 has been used in Bessa and Rodrigues (2016) to show the existence of *Cocoon bifurcations* in conservative systems. The tangencies of invariant manifolds coexist with transverse intersections that generate the hyperbolic horseshoes.

Theorem 3.10 (Bykov (1999, 2000) and Labouriau and Rodrigues (2015)). C^2 *arbitrarily close to any X whose parameters P lies in $\mathfrak{D}$, there is a vector field in $\mathfrak{B}$, where the non-hyperbolic dynamics cannot be separated by an isotopy from the maximal hyperbolic set that appears in any tubular neighborhood of the cycle.*

According to Theorem 3.5, there is an open set $\mathfrak{U} \subset \mathfrak{B}$ where, densely, the two-dimensional invariant manifolds of two hyperbolic saddle-foci O_1 and O_2 meet tangentially. From now on, we will make use of the following additional hypothesis:

(BC4) $\delta_1 = \alpha_1/\mu_1 > 1$ and $\delta_2 = \mu_2/\alpha_2 > 1$.

Property **(BC4)** implies that the first return map to $\text{In}(O_1)$, when well defined, is dissipative (determinant of $D\Pi$ less than one). It is required to prove the existence of strange attractors in the unfolding of the heteroclinic tangencies in Theorem 3.5. An explicit example with a two parametric vector field satisfying **(BC1)**–**(BC2)** and $\overline{\textbf{(BC3)}}$–**(BC4)** has been given in Labouriau and Rodrigues (2015).

The following technical lemma is the key to link Lemma 3.9 with strange attractors:

Lemma 3.11 (Labouriau and Rodrigues (2017)). *Let X_μ be a one-parametric family of vector fields in $\mathfrak{B}$ under the extra assumption **(BC4)** unfolding generically a heteroclinic tangency between $W^u(O_2)$ and $W^s(O_1)$ for $\mu = 0$. Then there exists a sequence of real numbers $\mu_j \to 0$ for which the first return map to $In(O_1)$, has a homoclinic tangency associated with a dissipative hyperbolic periodic orbit.*

The idea of the proof is quite standard. Assume that X_μ unfolds generically at $\mu = 0$ a heteroclinic tangency between $W^u(O_2)$ and $W^s(O_1)$. Theorem 3.2 says that there is a suspended horseshoe Λ in a rectangle (say $R_1 \subset \text{In}(O_1)$) near the two-dimensional connection $O_2 \to O_1$. Within this rectangle, the local stable foliation associated with Λ is horizontal as $W^s(O_1) \cap \text{In}(O_1)$ and the local unstable foliation associated with Λ is vertical as $W^u(O_2) \cap \text{In}(O_1)$. The lines corresponding to the unstable manifolds of the periodic solutions of Λ are very close to $W^u(O_2) \cap \text{In}(O_1)$ and their first return map to $\text{In}(O_1)$ have the same shape as the line $\Pi(W^u(O_2) \cap \text{In}(O_1))$. Analogously, the lines corresponding to the stable manifolds of the periodic solutions of Λ are very close to $W^s(O_1) \cap \text{In}(O_1)$. Hence near the point of tangency between $W^u(O_2)$ and $W^s(O_1)$, there are infinitely many lines corresponding to the stable and unstable manifold of periodic orbits lying in Λ. In particular, near $\mu = \mu_0$, there is at least one pair of invariant manifolds of a periodic point of Λ meeting tangentially. The dissipativeness follows from Property **(BC4)**.

In particular, as consequence of the result described in §2.5 we get:

Corollary 3.12. *There is an open set $\mathfrak{U}$ of vector fields within $\mathfrak{B}$ and satisfying the extra assumption **(BC4)** such that arbitrarily close to any $X \in \mathfrak{U}$, there is a vector field whose flow exhibits exhibiting infinitely many sinks and persistent strange attractors near the cycle Γ.*

3.5   Breaking the one-dimensional connection

Since the transversality of the two-dimensional invariant manifolds $W^u(O_2)$ and $W^s(O_1)$ is persistent under small perturbations, a generic unfolding of a Bykov cycle is characterized by the break of the connection along the one-dimensional invariant manifolds, a phenomenon of codimension two (see Appendix A). When the connection along the one-dimensional invariant manifolds breaks, the horseshoes in Theorem 3.2 are destroyed and hence the same mechanisms explaining the genesis of strange attractors in the unfolding of Shilnikov cycles are valid to provide a source of sinks and strange attractors (see Theorem 2.17).

To be more precise, we consider a two-parameter family of differential equations

$$\dot{u} = X_\mu(u), \qquad u \in \mathbb{R}^3, \qquad \mu \in \mathbb{R}^2, \tag{3.12}$$

where the flow of (3.12) has a Bykov cycle for $\mu = (0,0)$ and the connection $\gamma_{1\to2}$ from O_1 to O_2 unfolds generically.

Theorem 3.13 (Bykov (1999, 2000)). *If $\|\mu\|$ is sufficiently small, then there exist two bifurcation curves ℓ_1 and ℓ_2 such that if $\mu \in \ell_j$, then the system (3.12) has a homoclinic curve associated with the saddle-focus O_j, $j = 1, 2$.*

Note that if Shilnikov condition **(S3b)** is met for the saddle-foci, then strange attractors appear.

The above result establishes a correspondence between tangencies in the bifurcation diagram and in the phase space. Each of the curves of Theorem 3.13 has a form of a spiral winding toward the origin. The lines ℓ_1 and ℓ_2 may be interpreted as $\Phi_2^{-1}(W_{\text{loc}}^s(O_1) \cap \text{Out}(O_2))$ and $\Psi_{1\to2} \circ \Phi_1(W_{\text{loc}}^u(O_2) \cap \text{In}(O_1))$.

As explored in Bykov (1999, 2000), deeper results may be stated with respect to the relative position of the curves ℓ_1 and ℓ_2. Defining G as

$$G = \frac{\alpha_1^2}{\mu_1^2\,\omega_1^2} + \frac{\mu_2^2}{\alpha_2^2\,\omega_2^2} - \frac{2\,\alpha_1\,\mu_2}{\mu_1\,\alpha_2\,\omega_1\,\omega_2}\mathcal{E} + (1 - \mathcal{E}^2),$$

where $\mathcal{E} = \frac{1}{2}\left(a^2 + \frac{1}{a^2}\right) \geq 1$, one gets:

Theorem 3.14 (Bykov (1999, 2000)). *For every generic two-parametric family as in (3.12), there exists a sequence $(\mu_j)_{j \in \mathbb{N}}$ of parameters, for which the flow of X_{μ_j} has the coexistence of two homoclinic cycles associated with O_1 and O_2. If $G > 0$, then at each point $\mu = \mu_j$, the curves intersect transversally. If $G < 0$, by any arbitrarily small perturbation of the family (3.12), one may obtain a tangency of ℓ_1 and ℓ_2 arbitrarily close to $\mu = 0$.*

Historical remark

Bykov never comments on the chiralities of the nodes, assuming implicitly that they satisfy either **(BC2)** or (**BC3**). Therefore, in the cross sections, the spirals corresponding to the two-dimensional invariant manifolds of the saddle-foci are oriented in the same way. This explains the orientation of the spirals of Figure 2 of Bykov (2000) in contrast to those shown in Figure 11 of Knobloch, Lamb, and Webster (2014), that turn in opposite directions because the nodes have the same chirality. Bykov's condition $G < 0$ in Theorem 3.2 of Bykov (2000) is analogous to our condition (3.11) that defines the set $\mathcal{A}$ where tangencies are dense.

4 *Bifocus homoclinic cycles*

The homoclinic cycle to a bifocus equilibrium provides one of the main examples of the occurrence of chaotic dynamics in four-dimensional vector fields. The striking complexity of the dynamics near homoclinic cycles has been discovered and investigated by Shilnikov under non-resonant condition in Shilnikov (1967, 1970). Namely, Shilnikov proved the existence of a countable set of periodic solutions of saddle type. Later the existence of three-dimensional horseshoes explaining geometrically the presence of this periodic obits was presented in Ibáñez and Rodrigues (2015) and Wiggins (2013).

The spiraling geometry of the non-wandering set near the homoclinic cycle associated with the non-resonant bifocus has been partially described in Fowler and Sparrow (1991) where the authors studied generic unfoldings of these cycles. On the other hand the resonant case includes Hamiltonian and reversible systems. The presence of two dimensional horseshoes in any neighborhood of a non-degenerate Hamiltonian bifocal homoclinic orbit was obtained by Devaney (1976a). This result was extended by citelerman1991complex,L97,lerman2000dynamical (see also Koltsova and Lerman (2009)) for nearby level sets. Also infinitely many secondary bifocal homoclinic orbits accompanying the primary connection have been find in the Hamiltonian case. In the reversible context this result was extended for non-degenerate symmetric bifocal homoclinic orbits by Härterich (1998). In fact, the similarity between reversible and Hamiltonian has been demonstrated in many cases Devancy (1976b). For instance, both reversible and Hamiltonian homoclinic orbits are accompanied by a one-parameter family of periodic orbits Devaney (1977). In the general context, the complete understanding of the structure of this spiraling set is a hard task.

In this chapter we will study the non-resonant configuration as well as the Hamiltonian and reversible setting. These resonant configurations appear in the limit family of generic unfoldings of nilpotent singularities while the non-resonant case emerges from these families into the unfolding as showed in Chapter 5. We will focus in explaining the aforementioned results and show how strange attractors, attracting invariant tori, robust heterodimensional cycles and robust tangencies can be obtained by perturbation these cycles following the recent works Barrientos, Ibáñez, and Rodríguez (2016), Barrientos, Raibekas, and Rodrigues (2019), and Rodrigues (2018).

4.1   Setting

Consider a differential equation

$$\dot{u} = X(u), \qquad u \in \mathbb{R}^4 \tag{4.1}$$

where X is a C^r vector field with $r > 1$ as large as necessary. We assume that the associated flow satisfies the following hypotheses:

(B1) There is an equilibrium point O with eigenvalues of $DX(O)$

$$-\alpha_1 \pm i\omega_1 \quad \text{and} \quad \alpha_2 \pm i\omega_2 \quad \text{where } -\alpha_1 < 0 < \alpha_2 \text{ and } \omega_1, \omega_2 > 0.$$

(B2) There is (at least) one non-degenerate homoclinic orbit γ. That is,

$$\gamma \subset W^s(O) \cap W^u(O)$$

and

$$\dim T_x W^s(O) \cap T_x W^u(O) = 1 \quad \text{for all } x \in \gamma.$$

The homoclinic cycle is given by $\Gamma = \{O\} \cup \gamma$. The equilibrium O possesses two-dimensional stable and unstable manifolds, $W^s(O)$ and $W^u(O)$ respectively, which intersect non-transversely along γ. If we restrict the system to $W^s(O)$, the equilibrium O is a stable focus, i.e., the orbits spiral around O as $t \to \infty$. Similar behavior occurs when we restrict the system to $W^u(O)$. For this reason O is called *bifocus*. Similarly Γ is called non-degenerate *bifocus homoclinic cycle*. Sometimes in the literature this configuration has also been referred as non-degenerate *bifocal homoclinic cycle* (or orbit).

We will consider different conditions on the *saddle index*

$$\delta = \frac{\alpha_1}{\alpha_2} > 0.$$

Namely, we study the above configuration under the assumptions

(B3) $\delta \neq 1$ (Shilnikov condition)

(B4) $\delta = 1$ (resonant condition).

Shilnikov (1967) discovered that, when **(B3)** holds, the dynamics near to Γ involves infinitely many periodic solutions arbitrarily close to the cycle. For this reason **(B3)** is usually called *Shilnikov condition* and when Γ satisfy this condition is called non-degenerate *Shilnikov bifocus homoclinic cycle*. On the other hand, the resonant case **(B4)** includes Hamiltonian and reversible systems. Devaney (1976a) and Devaney (1977) (see also Belyakov (1984a), Härterich (1998), and Lerman (1991)) showed that in these resonant cases the γ is also accompanied by infinitely many periodic orbits and moreover, infinitely many secondary homoclinic orbits. Throughout the present chapter these results will be clarified.

4.2 Local map: spiraling geometry

Since X is a C^r vector field with $r > 1$ large enough, under non-resonant conditions on the eigenvalues, we can get a linearize (4.1) around O as smoothly as necessary (see §1.1). That is, there exists a neighborhood V of O such that, if (x_1, x_2, y_1, y_2) is in V, the system (4.1) is orbitally equivalent to the linear system

$$\begin{cases} \dot{x}_1 = -\alpha_1 x_1 - \omega_1 x_2 \\ \dot{x}_2 = \omega_1 x_1 - \alpha_1 x_2 \\ \dot{y}_1 = \alpha_2 y_1 - \omega_2 y_2 \\ \dot{y}_2 = \omega_2 y_1 + \alpha_2 y_2. \end{cases} \tag{4.2}$$

Using (4.2), we may define bipolar coordinates $(r_s, \theta_s, r_u, \theta_u)$ near the bifocus O. Namely,

$$x_1 = r_s \cos \theta_s \quad x_2 = r_s \sin \theta_s \qquad \text{and} \qquad y_1 = r_u \cos \theta_u \quad y_2 = r_u \sin \theta_u.$$

The local invariant manifolds are given by

$$W^s_{\text{loc}}(O) = \{r_u = 0\} \qquad \text{and} \qquad W^u_{\text{loc}}(O) = \{r_s = 0\}.$$

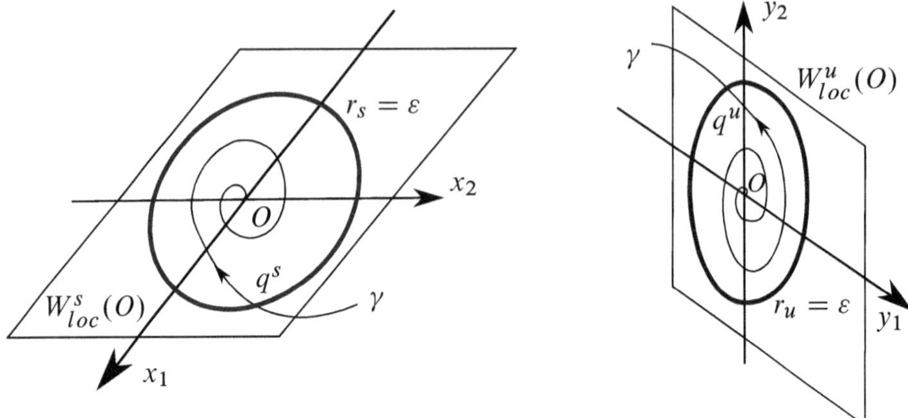

Figure 4.1: *Coordinates near O.*

Near O, in bipolar coordinates $(r_s, \theta_s, r_u, \theta_u)$, the dynamics is governed by the differential equations

$$\dot{r}_s = -\alpha_1 r_s \qquad \dot{\theta}_s = \omega_1 \qquad \dot{r}_u = \alpha_2 r_u \qquad \dot{\theta}_u = \omega_2.$$

whose explicit solutions starting at a point $(r_s, \theta_s, r_u, \theta_u)$ is given by

$$r_s(t) = r_s e^{-\alpha_1 t} \qquad \theta_s(t) = \theta_s + \omega_1 t \qquad r_u(t) = r_u e^{\alpha_2 t} \qquad \theta_u(t) = \theta_u + \omega_2 t.$$

As illustrated in Figure 4.2, in order to construct the local return map we consider two sections near the origin, Σ^s and Σ^u, which are solid tori defined by

$$\Sigma^s = \{r_s = \varepsilon\} \qquad \text{and} \qquad \Sigma^u = \{r_u = \varepsilon\} \tag{4.3}$$

where $\varepsilon > 0$ is chosen sufficiently small such that

$$\{q^s\} = \gamma \cap \Sigma^s \subset W^s_{\text{loc}}(O) \qquad \text{and} \qquad \{q^u\} = \gamma \cap \Sigma^u \subset W^u_{\text{loc}}(O).$$

Trajectories starting at Σ^s and Σ^u go outside of V in negative and positive time, respectively. For convenience, we write $(\theta_s, r_u, \theta_u)$ and $(r_s^*, \theta_s^*, \theta_u^*)$ for the coordinates in Σ^s and Σ^u respectively.

The time of flight inside V of a trajectory with initial condition $(\theta_s, r_u, \theta_u)$ in $\Sigma^s \setminus W^s_{\text{loc}}(O)$ is given by

$$\frac{1}{\alpha_2} \ln\left(\frac{\varepsilon}{r_u}\right).$$

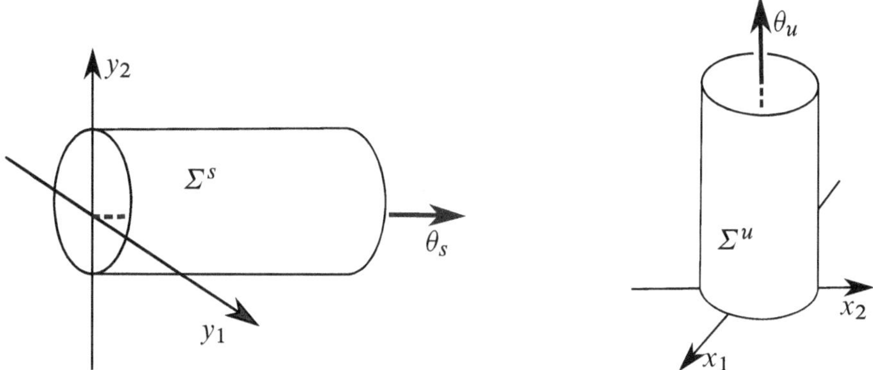

Figure 4.2: *Cross-sections near O.*

Recalling the variables, we may take $\varepsilon = 1$. Thus, we may define the map $\Phi : \Sigma^s \setminus W^s_{loc}(O) \to \Sigma^u$,

$$r_s^* = r_u^\delta \qquad \theta_s^* = \theta_s - \frac{\omega_1}{\alpha_2} \ln r_u \qquad \theta_u^* = \theta_u - \frac{\omega_2}{\alpha_2} \ln r_u. \qquad (4.4)$$

A similar calculation provides that the inverse map $\Phi^{-1} : \Sigma^u \setminus W^u_{loc}(O) \to \Sigma^s$ of the local map Φ is given by

$$\theta_s = \theta_s^* + \frac{\omega_1}{\alpha_1} \ln r_s^* \qquad r_u = (r_s^*)^{1/\delta} \qquad \theta_u = \theta_u^* + \frac{\omega_2}{\alpha_1} \ln r_s^*. \qquad (4.5)$$

Spiralling geometry Now we describe the spiraling behavior of solutions near O. We address the reader to Definition 2.2 to recall the notion of spiral. Additional definitions are needed:

Definition 4.1. *A two-dimensional manifold S embedded in $\mathbb{R}^3$ is called a* spiraling sheet *accumulating on a curve C if there are neighborhoods $U \subset \mathbb{R}^3$, $W \subset \mathbb{R}^2$ of C and $(0,0)$ respectively, a spiral $S \subset \mathbb{R}^2$ around the origin, a non-degenerate closed interval I and a diffeomorphism $\eta : U \to I \times W$ such that*

$$\eta(S \cap U) = I \times (S \cap W) \quad and \quad C = \eta^{-1}(I \times \{0\}).$$

The curve C is also called the basis *of the spiraling sheet.*

Figure 4.3: *Spiralling sheet*

Up to a diffeomorphism, we may think on a spiraling sheet accumulating on a curve as the cartesian product of a spiral and a curve. In fact, concerning Definition 4.1, we will consider that the curve C lies on the invariant manifolds of O, when restricted to the above cross sections. Each cross-section to C intersects the spiraling sheet S into a spiral. Note also that the diffeomorphic image of a spiraling sheet contains a spiraling sheet.

Definition 4.2. *Given two spiraling sheets S_1 and S_2 accumulating on the same curve $C \subset \mathbb{R}^3$, any region limited by S_1 and S_2 inside a tubular neighborhood of C is said a* scroll *accumulating on C.*

The following result will be essential in the sequel. It shows that a set diffeomorphic to a disc transverse to $W^s_{loc}(O) \cap \Sigma^s$ is sent by Φ into a spiraling sheet.

Proposition 4.3 (Härterich (1998) and Ibáñez and Rodrigues (2015)). *For $v > 0$ arbitrarily small, let $\Xi : D \subset \mathbb{R}^2 \to \mathbb{R}$ be a C^1 map defined on the disc $D = \{(u,v) \in \mathbb{R}^2 : 0 \leqslant u^2 + v^2 \leqslant v < 1\}$ and let*

$$\mathcal{F} = \{(\theta_s, r_u, \theta_u) \in \Sigma^s : \theta_s = \Xi(r_u \cos \theta_u, r_u \sin \theta_u), \ 0 \leqslant r_u \leqslant v\}$$

and

$$\mathcal{F}^* = \{(r_s^*, \theta_s^*, \theta_u^*) \in \Sigma^u : \theta_u^* = \Xi(r_s^* \cos \theta_u^*, r_s^* \sin \theta_u^*), \ 0 \leqslant r_s^* \leqslant v\}.$$

Then the sets $\Phi(\mathcal{F} \setminus W^s_{loc}(O))$ and $\Phi^{-1}(\mathcal{F}^* \setminus W^u_{loc}(O))$ are spiraling sheets accumulating on $W^u_{loc}(O) \cap \Sigma^u$ and $W^s_{loc}(O) \cap \Sigma^s$ respectively.

As a consequence of the above proposition it follows that:

Remark 4.4. *The image or the pre-image by Φ of any cylindrical neighborhood C of a segment in $W^s_{loc}(O) \cap \Sigma^s$ or $W^u_{loc}(O) \cap \Sigma^u$ is a scroll in Σ^u or Σ^s respectively.*

We observe the duality in the spiraling geometry behavior showed in the previous proposition. This follows from the relative symmetry of Σ^s and Σ^u and the expressions (4.4) and (4.5) for Φ and Φ^{-1} respectively.

4.3 Shilnikov bifocus homoclinic cycles

Now we will consider the vector field X in (4.1) under the assumptions **(B1)**–**(B3)**. In order to describe the dynamics near the homoclinic cycle Γ we will need first to characterize the global map.

Global map. Recall that γ intersects Σ^s and Σ^u at q^s and q^u respectively. Without loss of generality we can assume that $q^s = (0,0,0)$ and $q^u = (0,0,0)$ in the coordinates of the corresponding cross-sections. Therefore, we can choose two neighborhoods $V^s \subset \Sigma^s$ and $V^u \subset \Sigma^u$ of q^s and q^u respectively such that the map $\Psi : V^u \to V^s$ generated by the global flow near Γ is a diffeomorphism. Now we define local rectangular coordinates:

$$y_1 = r_u \cos(\theta_u) \qquad y_2 = r_u \sin(\theta_u) \qquad z_s = \theta_s \qquad \text{in} \quad \Sigma^s$$

and

$$x_1 = r_s^* \cos(\theta_s^*) \qquad x_2 = r_s^* \sin(\theta_s^*) \qquad z_u = \theta_u^* \qquad \text{in} \quad \Sigma^u.$$

Following Fowler and Sparrow (1991), the global map Ψ could be approximated by the linear transformation

$$\begin{pmatrix} y_1 \\ y_2 \\ z_s \end{pmatrix} = A \begin{pmatrix} x_1 \\ x_2 \\ z_u \end{pmatrix}$$

where A is a invertible matrix. Moreover, according to the non-degenerate condition in **(B2)** this matrix must to satisfy that v and Av are non-collinear vectors where $v = (0,0,1)^T$. That is, v cannot be an eigenvector of A.

First-return map. In order to provide a well defined first-return map, we consider a cylindrical neighborhood

$$C^u = \{(r_s^*, \theta_s^*, \theta_u^*) \in V^u : |\theta_u^*| \leq \varepsilon, \; r_u \leq \kappa\} \subset \Sigma^u$$

of q^u for some small enough constant $\varepsilon, \kappa > 0$. Then we consider the set

$$S = \Phi^{-1}(C^u \setminus W^u_{loc}(O)) \subset \Sigma^s$$

and define the first return map as $\Pi : S \to \Sigma^s$ by $\Pi = \Psi \circ \Phi$. Observe that according to Remark 4.4 the set S is a scroll on Σ^s.

4.3.1 Horseshoes

Now we will show the existence of suspended horseshoes for a vector field under the assumptions **(B1)**–**(B3)**. In particular, we have two possibilities for hypothesis (B3):

(B3a) $\delta > 1$ (negative divergence),

(B3b) $\delta < 1$ (positive divergence).

The relative symmetry of the cross sections as well as the local and the global maps and their inverses allow us to get a similar first-return map around Γ for the vector field induced by reversing time. Thus, we may reduce **(B3b)** from **(B3a)** and vice versa. To be more specific, consider that Π is the first-return map of the vector field X under the assumption **(B3b)**. We observe that $\Pi^{-1} = \Phi^{-1} \circ \Psi^{-1} = \Psi \circ \widetilde{\Pi} \circ \Psi^{-1}$ where $\widetilde{\Pi} = \Psi^{-1} \circ \Phi^{-1}$ is the first-return map defined on Σ^u for the vector field $-X$. Since the saddle index of the equilibrium point O for the vector field X is $\delta < 1$ (assumption **(B3b)**) then the saddle index for $-X$ is $\delta > 1$. Moreover, the analytic expression of $\widetilde{\Pi}$ is similar as the first-return map of X but with $\delta > 1$ (assumption **(B3a)**). Thus, the result proved for $\delta > 1$ holds for $\widetilde{\Pi}$ and consequently, by the conjugation for Π^{-1}.

Given $\eta \in \mathbb{S}^1$, for k large enough, consider

$$C^u_k = \{(r^*_s, \theta^*_s, \theta^*_u) \in C^u : a_{k+1} \leqslant r^*_s \leqslant a_k\} \subset \Sigma^u$$

where

$$a_k = e^{\frac{-\alpha_1}{\omega_2}(\eta + 2\pi k)}. \tag{4.6}$$

Hence, from (4.5) we have

$$S_k = \Phi^{-1}(C^u_k) \subset \{(\theta_s, r_u, \theta_u) \in \Sigma^s : b_{k+1} \leqslant r_u \leqslant b_k\}$$

where

$$b_k = (a_k)^{1/\delta} = e^{\frac{-\alpha_2}{\omega_2}(\eta + 2\pi k)}. \tag{4.7}$$

Observe S is the union of the sets S_k. Moreover, when $\delta > 1$ (assumption **(B3a)**) is easy to check that $b_{k+1} > a_k$ for any k large enough. Similarly, when $\delta < 1$ (assumption **(B3b)**) then $a_{k+1} > b_k$ for all k large enough.

The next results gives the existence of suspended horseshoes in the neighbourhood of Γ.

Theorem 4.5. *Let $\Pi : S \to \Sigma^s$ be the first-return map of a vector field under the assumptions **(B1)–(B3)**. Then there is a sequence of hyperbolic compact invariant sets Λ_k in S of Π such that the restriction of Π to Λ_k is topologically conjugate to the full shift on two symbols.*

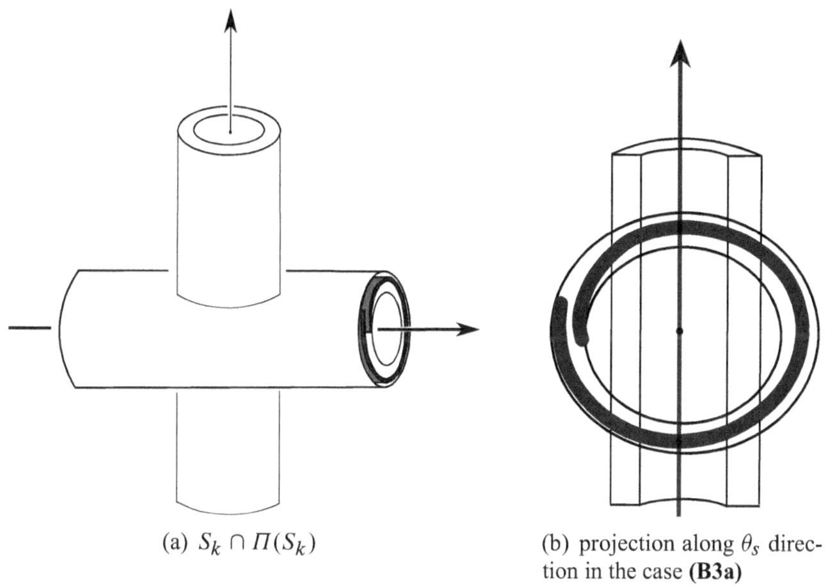

(a) $S_k \cap \Pi(S_k)$

(b) projection along θ_s direction in the case **(B3a)**

Figure 4.4: *Horseshoe.*

Proof. As we have remarked above, we can reduce **(B3b)** to **(B3a)** and vice versa. Hence, in order to explain the construction of the horseshoes we assume $\delta > 1$ (condition **(B3a)**).

Having into account that the map Ψ adds only a bounded distortion factor to the picture induced by the local map Φ we can assume that $\Psi(C_k^u)$ and C_k^u are of the same order. Thus, in view that under the assumption **(B3a)** one has that $b_{k+1} > a_k$ for k large enough and by means of an appropriate choice η to rotate S_k, we get that S_k and $\Pi(S_k) = \Psi(C_k^u)$ intersect as Figure 4.4 shows. The lateral view (projection along θ_s direction) of this intersection may seem similar to the horseshoes finding in the Shilnikov picture studied Chapter 2. In particular, $S_k \cap \Pi(S_k)$ consists in two component V_k^1 and V_k^2 which could be seen as a vertical slab as in the Smale horseshoe picture. On the other hand, $H_k^1 = \Pi^{-1}(V_k^1)$ and $H_k^2 = \Pi^{-1}(V_k^2)$ are those horizontal which fully intersect V_k^1 and V_k^2. An analysis of these intersections is done in Ibáñez and Rodrigues (2015) and Rodrigues (2018) (see also Shilnikov (1967) and Wiggins (2013)) where it is rigorously proved that the restriction of Π to the maximal invariant set Λ_k in S_k is topologically conjugate to the full shift of two symbols. The hyperbolicity of this set follows from Shilnikov (1970) (see Afraimovich et al. (2014)). $\qquad\square$

The effects of the complicate local map are reflected along the direction θ_s and θ_u. Namely, there are enormous stretching (in the θ_u direction) and contraction (in the θ_s direction). The radial direction r_u could be consider as the neutral direction in comparison with the two previous directions. When $\delta > 1$ (assumption **(B3a)**) we see contraction and while if $\delta < 1$ (assumption **(B3a)**), we have expansion. Fowler and Sparrow (1991, Sec. 5.1) provide the following asymptotic estimates[1] for the eigenvalues λ_s, λ_c and λ_u of the *full* first return-map Π:

$$|\lambda_s| = O(a_k) \ll 1 \qquad |\lambda_c| = O(1) \qquad |\lambda_u| = O(\frac{1}{b_n}) \gg 1 \qquad (4.8)$$

where a_k and b_n are the constant given above in (4.6) and (4.7). Moreover, as commented above, if $\delta > 1$ (assumption **(B3a)**) then

$$|\lambda_s| < |\lambda_c| < 1 < |\lambda_u|$$

and if $\delta < 1$ (assumption **(B3b)**) then

$$|\lambda_s| < 1 < |\lambda_c| < |\lambda_u|.$$

Figure 4.5 illustrates the stable and unstable manifolds of a periodic point of Π in a horseshoe Λ_k.

Linked horseshoes   Similar as in the case of Shilnikov homoclinic cycles studied in Chapter 2, the horseshoes Λ_k, Λ_{k+n} can be heteroclinically related. More precisely, for any n, we can find k such that $\Pi(S_i) \cap S_j$ for all $i, j \in \{k, \dots, k+n\}$. By a similar argument (see Ibáñez and Rodrigues (2015), Rodrigues (2018), and Shilnikov (1970)) the maximal invariant set in the union of S_i for $i = 1, \dots, k+n$ is a horseshoe in $N = 2n$ symbols. In particular, we also get sequence of hyperbolic compact invariant sets Ω_k accumulating on the cycle $\Gamma \cap \Sigma^s$ such that Π restrict to Ω_k is conjugate to a full shift on k symbols. Moreover, according to Shilnikov (1970) and Deng (1989a, 1993) (see Afraimovich et al. (2014)) similar symbolic description as in Theorem 2.8 can be also applied here.

---

[1] In the notation of Fowler and Sparrow (1991) $\varepsilon_1 = a_k$ for the estimate of $|\lambda_s|$ and $\varepsilon_1 = \kappa$, $\varepsilon_2 = \varepsilon, h = 1$ and $\bar{r}_2 = b_n$ for the estimation of $|\lambda_u|$.

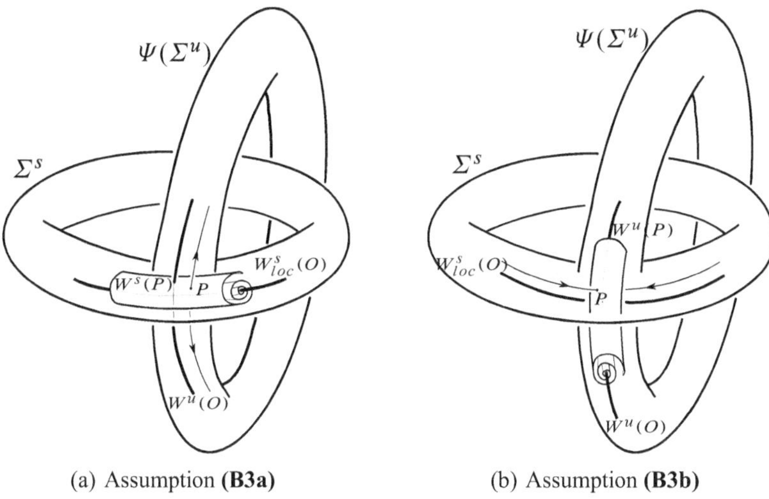

(a) Assumption (B3a)   (b) Assumption (B3b)

Figure 4.5: *Invariant manifold of a periodic point for the return map.*

Remark 4.6. *The periodic points in the horseshoes Ω_k are homoclinically related. That is, for any pair of periodic points P and Q in Ω_k an Ω_n respectively, $W^s(P)$ intersects transversally $W^u(Q)$ and vice versa. In particular, all the horseshoes belong to the same homoclinic class. Recall that the homoclinic class of a hyperbolic periodic point P is defined as the closure of hyperbolic periodic points Q whose stable and unstable manifolds intersect cyclically with those of P.*

4.3.2   Nearby Tatjer homoclinic tangencies

An important open question related to bifocus homoclinic cycles is what type of dynamics could be appeared after breaking the connection. We are particularly interested in the occurrence of strange attractors and attracting invariant tori. Under extra assumptions we will show that Tatjer homoclinic tangencies S. V. Gonchenko, V. S. Gonchenko, and Tatjer (2007) and Tatjer (2001) associated with dissipative saddles of the first-return map could be appeared for small perturbations of the vector field which, in principle, no longer have the original homoclinic cycle. As we will explain later, unfolding generically these tangencies introduced by Tatjer (2001), one concludes the presence of strange attractors and attracting invariant tori.

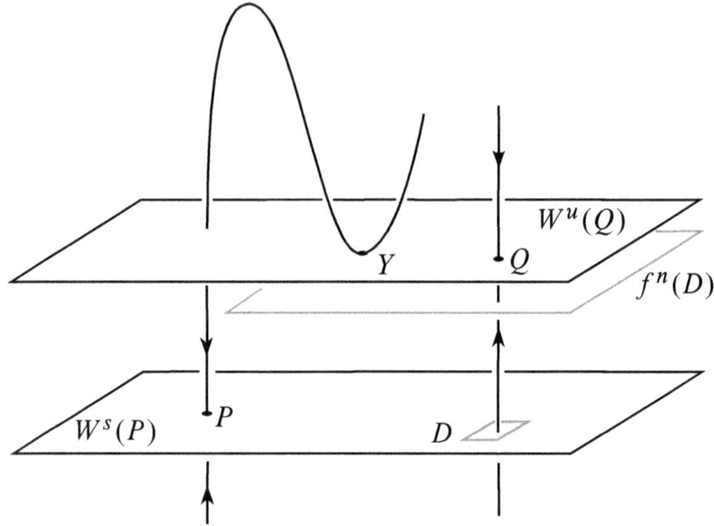

Figure 4.6: *Non-transverse equidimensional cycle.*

Creation of non-transverse equidimensional cycles.　First of all, we will explain how homoclinic tangencies appear after an arbitrarily small perturbation of the first-return map. In fact, these tangencies will be obtained from a kind of heteroclinic tangencies, the so-called non-transverse equidimensional cycles. Namely, we will say that a diffeomorphism f has a *non-transverse equidimensional cycle* if there are two homoclinically related periodic saddles P and Q having a heterodimensional tangency Y. See Figure 4.6.

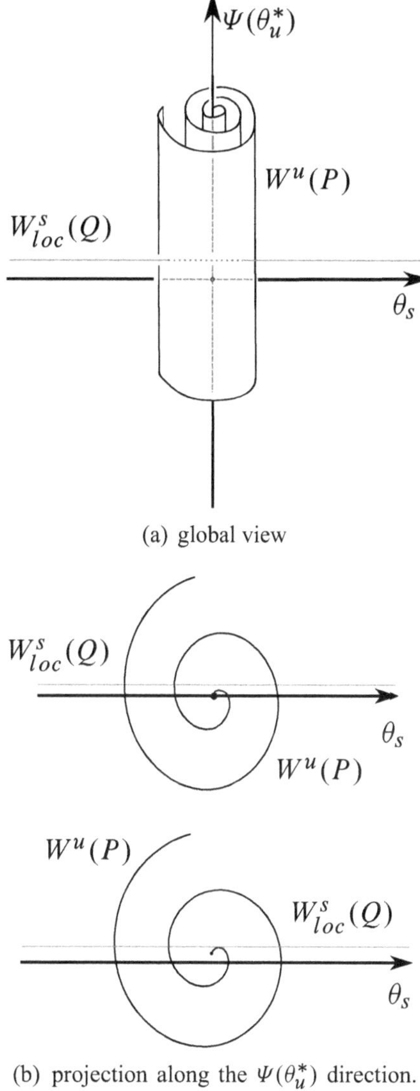

(a) global view

(b) projection along the $\Psi(\theta_u^*)$ direction.
Before and after the perturbation

Figure 4.7: *Unstable manifold of P and stable manifold of Q.*

Proposition 4.7 (Rodrigues (2018)). *Let $\Pi : S \to \Sigma^s$ be the first-return map of a vector field satisfying the hypotheses (B1)–(B3). Then Π can be C^r approximated by diffeomorphisms having a non-transverse equidimensional cycle associated with periodic points in the horseshoes.*

Proof. We assume the case $\delta < 1$ (hypothesis **(B3b)**) to explain the construction. The case $\delta > 1$ is analogous. Notice that horseshoes Ω_k accumulates on $\Gamma \cap \Sigma^s$ as well as the invariant manifolds of its periodic points. In particular, the local unstable manifold of one of this periodic points, say P, may be seen as a two-dimensional disc crossing transversally $W^s_{loc}(O) \cap V^s$. Thus, using Proposition 4.3, the set $\Pi(W^s_{loc}(P))$ is a spiraling sheet accumulating on $\Psi(W^u_{loc}(O) \cap V^u)$. On the other hand, the local stable manifold of a saddle point in Ω_k accumulates on $W^s_{loc}(O) \cap V^s$. Thus, by an arbitrarily small perturbation breaking the homoclinic connection we find a tangency between the $W^s_{loc}(P)$ and some saddle point Q. Indeed, this kind of perturbation unfolds the quasi-transverse intersection between $W^s_{loc}(O) \cap V^s$ and $\Psi(W^u_{loc}(O) \cap V^u)$ in Σ^s and hence we find a point Q in some horseshoe Ω_k for k large enough so that for a small perturbation $W^u(P)$ and $W^s(Q)$ has a tangency. See figure 4.7. $\square$

As a consequence of the above proposition we get easily a homoclinic tangency. To see this, observing Figure 4.6 we only need to apply λ-lemma to a small disc D in $W^u(P)$ containing the transversal intersection between $W^s(Q)$ and $W^u(P)$. The iterates of this disc will be close enough to $W^u(Q)$ and then after an arbitrarily small C^r perturbation of the homoclinic tangency Y we can perform a homoclinic tangency associated with P obtaining the following:

Corollary 4.8. *Let $\Pi : S \to \Sigma^s$ be the first-return map of a vector field under the hypotheses **(B1)**–**(B3)**. Then Π can be C^r approximated by diffeomorphisms having a homoclinic tangency associated with a periodic point in a horseshoes.*

We want to remark the following important fact:

Remark 4.9. *The above results hold for generic one-parameter unfoldings of the homoclinic orbit. Namely, in this case one gets a sequence of parameters μ_k converging to $\mu = 0$ (corresponding to the cycle) such that homoclinic tangencies (non-transverse equidimensional cycles) are obtained for $\mu = \mu_k$.*

However, these homoclinic tangencies are not enough to conclude the existence of Newhouse phenomenon and persistent strange attractors from its generic unfoldings as we did in §2.5. For diffeomorphisms in dimension three, the existence of these reach dynamics from a generic unfolding of a homoclinic tangency associated with a periodic saddle P has only been showed under extra assumptions. The sectional dissipativeness of the saddle point P is one of them (see S. V. Gonchenko, Turaev, and Shilnikov (1993), Palis and Viana (1994), and Viana (1993)). That is, if λ_1, λ_2 and λ_3 are the eigenvalues of $Df(P)$ then it requires that $|\lambda_i \lambda_j| < 1$ for all $i, j \in \{1, 2, 3\}$. However, in view of (4.8), we are not under this condition. Another condition was introduced by Tatjer in Tatjer (2001) (see also S. V. Gonchenko, V. S. Gonchenko, and Tatjer (2007)). In order to provide a rigorous definition we need some preliminary results.

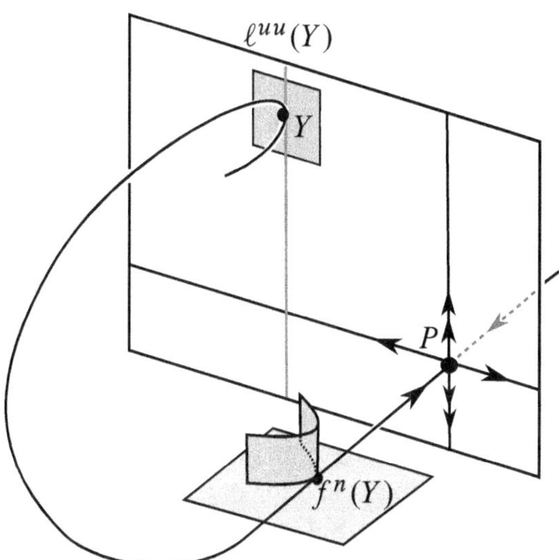

Figure 4.8: *Tatjer homoclinic tangency (type I)*

Tatjer homoclinic tangencies. Let P be a hyperbolic saddle fixed point of a three dimensional diffeomorphism f. For simplicity of the exposition we have chosen a fixed point but all the terminology and concepts are valid if P is a periodic point. Suppose that the $Df(P)$ has real eigenvalues λ_s, λ_{cu} and λ_{uu} satisfying the

$$|\lambda_s| < 1 < |\lambda_{cu}| < |\lambda_{uu}|.$$

Thus the tangent space at P has a dominated splitting of the form $E^s \oplus E^{cu} \oplus E^{uu}$ given by the corresponding eigenspaces. The unstable manifold $W^u(P)$ is tangent at P to the bundle $E^u = E^{cu} \oplus E^{uu}$. On the other hand, according to Hirsch, Pugh, and Shub (1977), the extremal bundle E^{uu} can be also integrated providing a one-dimensional manifold $W^{uu}(P)$ called strong unstable manifold. Moreover, this bundle can be uniquely extend to $W^u(P)$ providing a foliation $\mathcal{F}^{uu}(P)$ of this manifold by one-dimensional leafs $\ell^{uu}(Y)$ containing $Y \in W^u(P)$. We assume additionally that the center-stable bundle $E^s \oplus E^{cu}$ is also extended and integrated along the stable manifold $W^s(P)$ of P. Although the extended center-stable bundle is not unique any center-stable manifold contains $W^s(P)$ and any two of these manifolds are tangent to each other at every point of $W^u(P)$.

Definition 4.10. *A three-dimensional diffeomorphism as above has a* Tatjer homoclinic tangency *associated with P (which corresponds to the type I in Tatjer (2001)) if*

(T1) $W^s(P)$ and $W^u(P)$ have a quadratic tangency at Y which does not belongs to the strong unstable manifold $W^{uu}(P)$ of P,

(T2) $W^s(P)$ is tangent to the leaf $\ell^{uu}(Y)$ of $\mathcal{F}^{uu}(P)$ at Y,

(T3) $W^u(P)$ is transverse to center-stable manifold at Y.

Remark 4.11. *If P has stable index (dimension of the stable bundle) equals to two, the above definition of Tatjer homoclinic tangency applies to f^{-1}.*

In figure 4.8 we illustrate the condition (T3) around the point $\bar{Y} = f^n(Y) \in W^s_{loc}(P)$ for $n > 0$ large enough. At this point, condition (T3) can be read as the transversality between $W^u(P)$ and the surface S corresponding to the center-stable manifold tangent to $E^s \oplus E^{cu}$ at $\bar{Y}$. We must also notify that originally Tatjer (2001) includes the extra assumption of C^1 linearazing coordinates around P. Later in S. V. Gonchenko, V. S. Gonchenko, and Tatjer (2007) the results in Tatjer (2001) were generalized without this assumption.

Remark 4.12. *The conditions (T1) and (T3) are generic. This means that by an arbitrarily small perturbation one can always assume that a homoclinic tangency under the assumption (T2) is, in fact, a Tatjer tangency (type I).*

Although (T1) is a codimension one condition, we must observe that tangency requires in (T2) is a condition of codimension

$$\dim \mathbb{R}^3 - \dim[T_Y W^u(P) + T_Y \ell^{uu}(Y)] = 2.$$

For more details about tangencies of large codimension see also Barrientos and Raibekas (2017).

The next theorem summarized some of the main results in S. V. Gonchenko, V. S. Gonchenko, and Tatjer (2007) and Tatjer (2001) associated with Tatjer homoclinic tangencies:

Theorem 4.13. *Let f be a three-dimensional C^r diffeomorphism with $r \geqslant 5$ which has a Tatjer homoclinic tangency associated with a saddle periodic point whose eigenvalues are λ_s, λ_c and λ_u with*

$$|\lambda_s| < 1 < |\lambda_u|, \quad |\lambda_s| < |\lambda_c| < |\lambda_u| \quad \text{and} \quad |\lambda_s \lambda_c \lambda_u| < 1 \ \text{(dissipativeness)}.$$

Assume that either

$$
\begin{array}{llll}
\textit{(Case A)} & |\lambda_c| < 1 & |\lambda_c \lambda_u| > 1 & |\lambda_s \lambda_u| < 1 \qquad (4.9) \\
\textit{(Case B)} & |\lambda_c| > 1. & & \qquad\qquad\qquad\quad (4.10)
\end{array}
$$

Then, for every two-parameter family f_μ unfolding the homoclinic tangency of f at $\mu = 0$ it holds that

1. *there is a sequence of the parameter values μ_n converging to $\mu = 0$ such that f_μ has an n-periodic smooth normally-hyperbolic attracting invariant circle for $\mu = \mu_n$ for n large enough.*

2. *there is a set E of parameter values accumulating on $\mu = 0$ and with positive Lebesgue measure for which f_μ has a Hénon-like strange attractor near the orbit of tangency.*

The role played by the saddle-node bifurcations in the two-dimensional scenario (see e.g. Yorke and Alligood (1983)) will be played in the above theorem by the Bogdanov–Takens bifurcation for three-dimensional diffeomorphisms Broer, Roussarie, and Simó (1996). For such bifurcation of periodic points, the corresponding spectrum has one unipotent eigenvalue (double eigenvalue equal to 1 with associated eigenspace of dimension 1). The attracting invariant circles in the above theorem are generated by this bifurcation. On the other hand, the persistent strange attractor are generated in the Case A (i.e., if (4.9) holds) applying Viana (1993) to the limit family of return maps

$$F_{a,b}(x, y, z) = (z, bz, a + y + z^2) \tag{4.11}$$

which is, essentially, the Hénon maps with two parameters. In fact, the computation of the above limit family obtained after a renormalization process is one of the main results in the works of Tatjer. In the Case B (i.e., if (4.10) holds) the limit family is

$$F_{a,b}(x, y, z) = (z, a + by + z^2, y). \tag{4.12}$$

In this case the strange attractors are obtained using again the Bogdanov–Takens bifurcation to find near a generic homoclinic tangencies of dissipative periodic points and then apply the results in Mora and Viana (1993) and Viana (1993).

We must point out that the strange attractors obtained above are topologically one-dimensional (one direction of expansivity). As we have comment, they arise from the limit return maps in (4.11) and (4.12) but using Viana (1993) and are essentially Hénon-like. In fact, since only in Case B, (4.10), we have a two-dimensional unstable manifold, this case should be the natural setting in which topologically two-dimensional strange attractors (two direction of expansivity) show up. However, as far as we know, no proof of the existence of strange attractors in this case was given. The existence of such attractors was shown in Viana (1997), but for a simpler case. See this reference for the a more rigorous definition of strange attractors with more than one expansive direction.

For a proof of the existence of two-dimensional strange attractors of $F_{a,b}$ in (4.12) it is very important to note that for each $(a, b) \in \mathbb{R}^2$, every point $(x, y, z) \in \mathbb{R}^3$ goes by one iteration of $F_{a,b}$ into the surface $C_{a,b} = \{(x, y, z) : y = a + bz + x^2\}$. Hence, it is enough to study the dynamics of $F_{a,b}$ on $C_{a,b}$.

Then, it is not difficult to check that the map $F_{a,b}$ restricted to $C_{a,b}$ is conjugate to the family of endomorphisms defined on $\mathbb{R}^2$ by

$$T_{a,b}(x, y) = (a + y^2, x + by).$$

The dynamical behavior of this family is rather complicated as it was numerically pointed out in Pumariño and Tatjer (2007). The attractors found for a large set of parameters seem to be two-dimensional strange attractors (at least, the sum of the Lyapunov exponents being positive). Moreover, in Pumariño and Tatjer (2006), a curve of parameters $(a(t), b(t))$ has been constructed in such a way that the respective map $T_{a(t),b(t)}$ has an invariant region in $\mathbb{R}^2$ which is homeomorphic to a triangle. In analogy with the well-known relationship between the quadratic map and the tent-map, authors in Pumariño, Rodríguez, Tatjer, et al. (2013) and Pumariño, Rodríguez, Tatjer, et al. (2014) introduced a family Λ_t of piecewise linear maps defined on the triangle of vertices $(0,0)$, $(1,1)$ and $(2,0)$. Namely, $\Lambda_t = A_t \circ \mathcal{F}$ where

$$A_t = \begin{pmatrix} t & t \\ t & -t \end{pmatrix} \qquad \text{and} \qquad \mathcal{F}(x, y) = \begin{cases} (x, y) & \text{if } x < 1, \\ (2 - x, y) & \text{if } x \geqslant 1. \end{cases}$$

In fact, Λ_t belongs to a large family called expanding baker maps which generalized to the plane the notion of tent-map. Observe that on $y = 0$ this family is reduced to

$$h_t(x) = \begin{cases} tx & \text{if } x < 1, \\ t(2 - x) & \text{if } x \geqslant 1 \end{cases}$$

which is a reparametrization of the tent-map on $[0, 2]$. The name of expanding baker map is given by its geometric behavior: first folding along the line $x = 1$ and after that expanding according to A_t. One of the main differences between Λ_t and the classical models of baker's transformations is that bakers use a knife to knead the bread while the bakers of Λ_t knead without using a knife.

Coming back to $T_{a(t),b(t)}$, the model Λ_t exhibits the same types of possible strange attractors observed numerically for its quadratic analogous: simply connected, non-simply connected and non-connected. In Pumariño, Rodríguez, Tatjer, et al. (2015) it was analytically proved that Λ_t has indeed a simply connected strange attractor for $2^{-1/2}(\sqrt{2} + 1)^{1/4} < t \leqslant 1$. Renormalization operators and the proof of the existence of any number of non-connected strange attractors were developed in Pumariño, Rodríguez, and Vigil (2017) and Pumariño, Rodríguez, and Vigil (2018). See also Pumariño, Rodríguez, and Vigil (2019) for a view of the current state of the subject.

Although the dynamics of the quadratic family $T_{a,b}$ is more complicated than the expanding baker maps (even in dimension one for ten-map) it seems natural to assume that $T_{a,b}$ presents persistent strange attractors. However, the existence of strange attractors has not been proved for $T_{a,b}$. We claim that the families of return maps unfolding Tatjer homoclinic tangencies given in (4.12) have persistent two-dimensional strange attractors. The proof of such a result needs to overcome the long bridge from the results of Benedicks and Carleson (1985) to those of Benedicks and Carleson (1991) and after that Mora and Viana (1993).

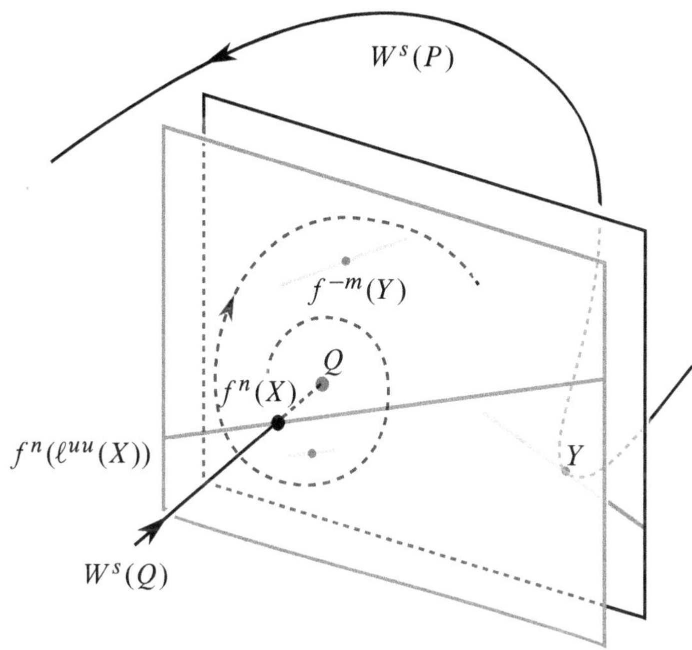

Figure 4.9: *Perturbation proceeded to create Tatjer tangencies.*

Tatjer tangencies from non-transverse equidimensional cycles. Recently, Kiriki, Nakano, and Soma (2017) have explained how it is possible to create Tatjer homoclinic tangencies from of a non-transverse equidimensional cycle with extra assumptions. Namely they assume that the periodic saddle P and Q involved in the equidimensional cycle has different signature. Namely, P has real eigenvalues while Q has a pair of non-real eigenvalues. The idea behind of the proof of the creation of a Tatjer tangency after arbitrarily small perturbation is showed in Figure 4.9. Basically consists again in to use the λ-lemma on a small disc D contained in the unstable manifold of $W^u(P)$ transverse at X to $W^s(Q)$. Now the presence of non-real eigenvalues of Q allows to $f^n(\ell^{uu}(X) \cap D)$

rotates around Q. Hence, perturbing if necessary the frequency of the complex eigenvalue to be irrational, the tangent space of $f^n(\ell^{uu}(X) \cap D)$ at $f^n(X)$ is arbitrarily close to the tangent space of $W^s(P)$ at $f^{-m}(Y)$ for $m > 0$ large enough. Then by a small C^r perturbation one gets a Tatjer homoclinic tangency associated with P (see Remark 4.12). Summarizing we have:

Theorem 4.14 (Kiriki, Nakano, and Soma (2017)). *Let f be a three-dimensional C^r diffeomorphisms having a non-transverse equidimensional cycle associated with periodic saddles P and Q where P has real eigenvalues and Q has a pair of non-real eigenvalues. Then f can be C^r approximated by diffeomorphisms having a Tatjer tangency associated with P.*

In order to apply this idea to the non-transverse equidimensional cycle obtained in Proposition 4.7 for a perturbation of the first-return map we will need the following assumption:

(B5) There exists a hyperbolic point Q with a pair of non-real eigen-values homoclinically related to a periodic point in one of the horseshoes.

Observe that this hypothesis could be obtained from a non-dominated context as it is proved in Bonatti, Díaz, and Pujals (2003, Lemm 1.9 and proof of Prop. 2.1) for homoclinic classes (see also Bessa, Rocha, and Varandas (2018, Thm. 2.8) where under stronger assumptions this type of results are revisited). Non-dominated dynamics is a natural assumption due to the plethora of bifurcations which arise when the cycle is broken. However, the main tool to go from non-dominated dynamics to periodic saddles with non-real eigenvalues in the above references is the Frank's lemma which only allows us to provide a C^1 approximation.

Theorem 4.15 (Rodrigues (2018)). *Let $\Pi : S \to \Sigma^s$ be the first-return map of a vector field under the assumptions **(B1)**–**(B3)** and **(B5)**. Then, Π can be C^r approximated by diffeomorphisms having a Tatjer tangency associated with a hyperbolic periodic point P in a horseshoe. Moreover, under the assumption **(B3a)** then the Tatjer homoclinic tangency is associated with a dissipative fixed point in the Case A of Theorem 4.13.*

Proof. Recall that all the periodic points in the horseshoes are homoclinically related (see Remark 4.6). On the other hand, since from the assumption **(B5)** the point Q with non-real eigenvalues values is homoclinically related with a periodic point in a horseshoe, by standard arguments using λ-lemma, we can assume that the non-transverse equidimensional cycle obtained in Proposition 4.7 is associated with the continuation of Q and a fixed point P in a horseshoe Λ_k

from Theorem 4.5 for k large enough. Now, Theorem 4.14 concludes that by an arbitrarily small C^r perturbation we obtain a Tatjer homoclinic tangency. Finally, from (4.8) and since $P \in \Lambda_k$, we conclude that the eigenvalues of P are λ_s, λ_c and λ_u with $|\lambda_s| = O(a_k) \ll 1$, $|\lambda_c| = O(1)$ and $|\lambda_u| = O(1/b_k) \gg 1$. Under the assumption **(B3b)**, we have that $|\lambda_c| < 1$ and $b_k > b_{k+1} > a_k$. Thus,

$$|\lambda_s \lambda_c \lambda_u| < 1 \qquad |\lambda_s \lambda_u| < 1 \qquad \text{and} \qquad |\lambda_c \lambda_u| > 1.$$

That is, P is a dissipative fixed point in the assumptions of the Case A of Theorem 4.13. This concludes the proof of the theorem. $\qquad\square$

As a consequence of the above theorem and Theorem 4.13, we get the following:

Corollary 4.16. *A vector field under the assumptions (B1), (B2), (B3a) and (B5) can be C^r approximated with $r \geqslant 5$ by vector fields with suspended Hénon-like strange attractors and attracting invariant tori.*

Some observations must be done. First, observe that in order to get attractors near of the bifocus homoclinic cycle seems necessary that the vector field has negative divergence (condition **(B3a)**) so that the flow near the equilibrium contracts volume. Under the condition **(B3b)**, the Tatjer homoclinic tangency is associated with a fixed point P of the first-return map Π which expands volume and has two-dimensional unstable manifold. Consequently, Case B in Theorem 4.13 is not possible. However, by considering Π^{-1}, the point P is now dissipative and it is not difficult to see that meets again the Tatjer conditions in Case A of Theorem 4.13. This implies the following remark:

Corollary 4.17. *A vector field under the assumptions (B1), (B2), (B3b) and (B5) can be C^r approximated with $r \geqslant 5$ by vector fields with suspended strange repellers and repelling tori.*

The second remark is concerned to the condition **(B5)**. This hypothesis is imposed on the initial vector field under the assumptions **(B1)**–**(B3)** in order to provide a readable explanation of the idea behind the construction of Tatjer homoclinic tangencies. One would expect to take out this condition by proving that periodic points with non-real eigenvalues appear after breaking the connection in the process of creation and destruction of infinitely many horseshoes. These points would be homoclinically related with the horseshoes that survive after the perturbation and one could expect to create a non-transverse equidimensional cycle with different signature after arbitrarily small perturbation.

4.4   Conservative bifocus homoclinic orbits

Now, we will consider that X in (4.1) is a Hamiltonian vector field. That is,

(BH) There exists a smooth function $H : \mathbb{R}^4 \to \mathbb{R}$ such that

$$X(v) = J \cdot \nabla H(v), \qquad v \in \mathbb{R}^4$$

where ∇H is the gradient of H and J is the Poisson matrix.

Under the assumption **(BH)**, the eigenvalues of O in **(B2)** are of the form

$$-\alpha \pm i\omega \qquad \alpha \pm i\omega \qquad \text{with} \quad \alpha\omega \neq 0.$$

Indeed, this follows from the fact that in this case $DX(O)$ is a Hamiltonian matrix and thus if λ is an eigenvalue, then so are $-\lambda, \bar{\lambda}$ and $-\bar{\lambda}$. Therefore, we are in the resonant case **(B4)**.

We will assume that (4.1) satisfies **(BH)**, **(B1)** and **(B2)**. Under these assumptions $\Gamma = \{O\} \cup \gamma$ is usually refer as (non-degenerate) *conservative bifocus homoclinic cycle*. By translating our coordinate frame, we can assume $H(O) = 0$. The two-dimensional invariant manifolds $W^s(O)$ and $W^u(O)$ are both contained in $H^{-1}(0)$. This level set is a smooth three-dimensional submanifold near every point, except for the bifocus equilibrium O, where it has a singularity. Hence, the stable and unstable manifolds of any bifocus equilibrium point generically intersect each other transversally in the level set. Thus we must to have the following remark:

Remark 4.18. *The assumption (B2) is actually satisfied for any generic Hamiltonian system with a bifocus equilibrium point.*

In this section, we will provide a symbolic description of the hyperbolic sets lying in a neighborhood of the non-degenerate homoclinic orbit γ. To this end, we will study the Poincaré return map through an appropriate cross-section.

4.4.1   The first-return map

Since X is a smooth Hamiltonian vector field, in order to describe the local behavior of the flow near the bifocus equilibrium, we can use the Moser's normal form (see Moser (1958) for the analytic case, Lyčagin (1977) for C^∞ vector fields or Banyaga, de la Llave, and Wayne (1996) and Bronstein and Kopanskii (1996) for some sufficiently smooth cases). These results guarantee that, in some neighborhood V of p, there exist local symplectic coordinates (x_1, x_2, y_1, y_2) such that the Hamiltonian takes the form

$$H(x_1, x_2, y_1, y_2) = h(\xi, \eta) = \alpha\xi + \omega\eta + \cdots$$

where

$$\xi = x_1 y_1 + x_2 y_2 \qquad \eta = x_1 y_2 - x_2 y_1$$

h is a smooth function and dots stand for higher order terms in ξ, η. We will work locally in these coordinates for which, in the neighborhood V, we have the following differential equations:

$$\begin{cases} \dot{x}_1 = -H_{y_1} = -h_\xi x_1 + h_\eta x_2 \\ \dot{x}_2 = -H_{y_2} = -h_\eta x_1 - h_\xi x_2 \\ \dot{y}_1 = H_{x_1} = h_\xi y_1 + h_\eta y_2 \\ \dot{y}_2 = H_{x_2} = -h_\eta y_1 + h_\xi y_2. \end{cases} \tag{4.13}$$

This system could be view as Hamiltonian analogous of the linear system (4.2). Introducing bipolar coordinates $(r_s, \theta_s, r_u, \theta_u)$, (4.13) can be written as

$$\dot{r}_s = -h_\xi r_s \qquad \dot{\theta}_s = -h_\eta \qquad \dot{r}_u = h_\xi r_u \qquad \dot{\theta}_u = -h_\eta. \tag{4.14}$$

In these bipolar coordinates invariant manifolds in V are linear. Namely,

Remark 4.19. *The stable (resp. unstable) manifold coincides with the stable (resp. unstable) subspace $r_u = 0$ (resp. $r_s = 0$). Moreover, the functions ξ and η are first integrals of the Hamiltonian vector field in (4.13).*

Cross-sections. Similar as §4.2 the solid tori

$$\Sigma^s = \{r_s = \varepsilon\} \quad \text{and} \quad \Sigma^u = \{r_u = \varepsilon\}$$

are cross-sections for the flow of (4.14) in V. Let us denote $\{q^s\} = \gamma \cap \Sigma^s$ and $\{q^u\} = \gamma \cap \Sigma^u$. Without loss of generality, we can assume that $q^s = (\varepsilon, 0, 0, 0)$ and $q^u = (0, 0, \varepsilon, 0)$. In order to avoid confusion we will denote by $(\theta_s, r_u, \theta_u)$ and $(r_s^*, \theta_s^*, \theta_u^*)$ the variables on Σ^s and Σ^u respectively.

On the other, according to Remark 4.19, since h_ξ and h_η are functions of ξ and η, they also remain constant along the orbits of (4.14) in V. Taking into account this observation, equations in (4.14) can be integrated. Following the calculation in §4.2, for an initial condition

$$x_{10} = r_{s0} \cos \theta_{s0} \quad x_{20} = r_{s0} \sin \theta_{s0} \quad y_{10} = r_{u0} \cos \theta_{u0} \quad \text{and} \quad y_{20} = r_{u0} \sin \theta_{u0}$$

we get the local map $\Phi : \Sigma^s \setminus W_{loc}^s(O) \to \Sigma^u$ given by

$$r_s^* = r_{u0} \qquad \theta_s^* = \theta_{s0} - \frac{\omega_0}{\alpha_0} \log \frac{\varepsilon}{r_{u0}} \qquad \theta_u^* = \theta_{u0} - \frac{\omega_0}{\alpha_0} \log \frac{\varepsilon}{r_{u0}} \tag{4.15}$$

where $\alpha_0 = h_\xi(\xi_0, \eta_0)$ and $\omega_0 = h_\eta(\xi_0, \eta_0)$ being

$$\xi_0 = x_{10}y_{10} + x_{20}y_{20} \quad \text{and} \quad \eta_0 = x_{10}y_{20} - x_{20}y_{10}.$$

We will use the local invariance of the functions ξ, η to introduce new coordinates (θ_s, ξ, η) and $(\xi^*, \eta^*, \theta_u^*)$ on Σ^s and Σ^u, respectively. These coordinates are given in the following way:

$$
\begin{aligned}
\xi &= \varepsilon r_u \cos(\theta_u - \theta_s) & \xi^* &= \varepsilon r_s^* \cos(\theta_u^* - \theta_s^*) \\
\eta &= \varepsilon r_u \sin(\theta_u - \theta_s) & \eta^* &= \varepsilon r_s^* \sin(\theta_u^* - \theta_s^*).
\end{aligned}
\tag{4.16}
$$

Thus, denoting $\Theta = \theta_u - \theta_s$, it holds that

$$\xi^2 + \eta^2 = (r_u \varepsilon)^2 \qquad \cos \Theta = \frac{\xi}{\sqrt{\xi^2 + \eta^2}} \qquad \text{and} \qquad \sin \Theta = \frac{\eta}{\sqrt{\xi^2 + \eta^2}}.$$

Note that $\Theta(\xi, \eta)$ is simply the polar angle of the point (ξ, η). Similar expressions follow for the coordinates on Σ^u. So, we obtain that

$$
\begin{aligned}
\Sigma^s &= \{(\theta_s, \xi, \eta) : \ \theta_s \in \mathbb{S}^1, \ |\xi|, |\eta| \leqslant \varepsilon\}, \\
\Sigma^u &= \{(\xi^*, \eta^*, \theta_u^*) : \ \theta_u^* \in \mathbb{S}^1, \ |\xi^*|, |\eta^*| \leqslant \varepsilon\}.
\end{aligned}
$$

Submanifolds Σ^s, Σ^u are foliated by levels $H = c$ into 2-dimensional annuli Σ_c^s, Σ_c^u, respectively. In the neighborhood V of p, one may regard equation $h(\xi, \eta) = c$ to be uniquely solved with respect to ξ,

$$\xi = a_c(\eta) = \alpha^{-1}c - \alpha^{-1}\omega\eta + \cdots$$

This allows us to replace ξ by a new coordinate c in each cross-section Σ^s and Σ^u. Thus, for each $c \in \mathbb{R}$ with $|c|$ small enough,

$$
\begin{aligned}
\Sigma_c^s &= \{(\theta_s, \eta) : \ \theta_s \in \mathbb{S}^1, \ |\eta| \leqslant \varepsilon\} \quad \text{and} \\
\Sigma_c^u &= \{(\theta_u^*, \eta^*) : \ \theta_u^* \in \mathbb{S}^1, \ |\eta^*| \leqslant \varepsilon\}.
\end{aligned}
$$

Remark 4.20. *The intersection of the stable (resp. unstable) manifold of p with Σ^s (resp. Σ^u) is given by $c = 0$ and $\eta = 0$ (resp. $\eta^* = 0$).*

Observe that the submanifolds Σ_c^s and Σ_c^u are symplectic ones with respect to a restriction of the differential form $\Omega = dx_1 \wedge dy_1 + dx_2 \wedge dy_2$. Moreover, since c is preserved by the flow, a straightforward calculation shows that restrictions of the form Ω to each annulus are given by 2-forms $d\theta_s \wedge d\eta$ and $d\theta_u^* \wedge d\eta^*$, respectively. Hence these new coordinates are symplectic on Σ_c^s and Σ_c^u. It means that, in particular, the local map and the global map restricted to the annuli Σ_c^s and Σ_c^u, respectively, are symplectic and hence both of them preserve area and orientation in these coordinates.

Local map. Next, we look for the expression of π^s restricted to the annuli Σ_c^s. From (4.15) and (4.16) we conclude that

$$\eta^* = \varepsilon r_{u0} \sin(\theta_{u0} - \theta_{s0}) = \eta_0. \tag{4.17}$$

Now, since $h(a_c(\eta), \eta) = c$, we get $a_c'(\eta) = -h_\eta(a_c(\eta), \eta)/h_\xi(a_c(\eta), \eta)$. In particular, evaluating at the initial point, it follows $a_c'(\eta_0) = -\omega_0/\alpha_0$ and also

$$r_{u0} = \varepsilon^{-1}\sqrt{a_c(\eta_0)^2 + \eta_0^2} \quad \text{and} \quad \theta_{u0} - \theta_{s0} = \Theta(a_c(\eta_0), \eta_0) \overset{\text{def}}{=} \Theta_c(\eta_0).$$

Thus, substituting into (4.15), we obtain

$$\theta_u^* = \theta_{s0} + a_c'(\eta_0) \log\left(\varepsilon^2 / \sqrt{a_c(\eta_0)^2 + \eta_0^2}\right) + \Theta_c(\eta_0). \tag{4.18}$$

Therefore, removing in (4.17) and (4.18) the subscript zero which indicates the evaluation at the initial point, we obtain the following expression of the local map Φ restricted to the annulus Σ_c^s:

$$\Phi_c : \Sigma_c^s \to \Sigma_c^u, \qquad (\theta_u^*, \eta^*) = \Phi_c(\theta_s, \eta) = (\theta_s + b_c(\eta), \eta) \tag{4.19}$$

where

$$b_c(\eta) = a_c'(\eta) \log\left(\varepsilon^2 / \sqrt{a_c(\eta)^2 + \eta^2}\right) + \Theta_c(\eta).$$

Note that $\Theta_c(\eta)$ is a bounded smooth function everywhere except at $\eta = 0$ for $c = 0$, where the lateral limits as $\eta \to 0$ are respectively $\Theta_0(0+) = \pi - \arctan(\alpha/\omega)$ and $\Theta_0(0-) = -\arctan(\alpha/\omega)$. The local map Φ_c is symplectic, discontinuous along the circle $\eta = 0$ for $c = 0$ and smooth for $c \neq 0$.

If $\theta_s = u(\eta)$ is a function defined for $|\eta|$ small enough, then the image by Φ_c of its graph is a curve in the annulus Σ_c^u, which is the graph of a function $\theta_u^* = u(\eta^*) + b_c(\eta^*) \bmod 2\pi$ with $\eta^* = \eta$.

Remark 4.21. *It follows from Lerman (1991, 1997, 2000) there is $\varepsilon > 0$ so that for $c \neq 0$ the map*

$$\varphi : [-\varepsilon, \varepsilon] \to \mathbb{R} \quad \text{given by} \quad \varphi(\eta) = u(\eta) + b_c(\eta)$$

is a unimodal function with critical point at

$$\eta_c = \frac{\omega^2 - \alpha^2}{\omega(\omega^2 + \alpha^2)} c + O(c^2)$$

and critical value

$$\varphi(\eta_c) = (\omega/\alpha) \log|c| + E(c)$$

where $E(c)$ is a bounded function

Global map.   Recall that γ intersects Σ^s and Σ^u at $q^s = (0,0,0)$ and $q^u = (0,0,0)$, respectively. Therefore, we can choose two neighborhoods $V^s \subset \Sigma^s$ and $V^u \subset \Sigma^u$ of q^s and q^u, respectively, such that the map $\Psi : V^u \to V^s$ generated by the global flow near γ is a diffeomorphism. This map is represented as a family of symplectic maps $\Psi_c : V_c^u \to V_c^s$ defined for every c with $|c|$ small enough and where $V_c^u = \Sigma_c^u \cap V^u$ and $V_c^s = \Sigma_c^s \cap V^s$. These symplectic diffeomorphisms have the form

$$\Psi_c(\theta_u^*, \eta^*) = (P_c(\theta_u^*, \eta^*), Q_c(\theta_u^*, \eta^*)), \quad \text{with} \quad \det \frac{D(P_c, Q_c)}{D(\theta_u^*, \eta^*)} \equiv 1 \tag{4.20}$$

and where P_c and Q_c are smooth functions. According to Remark 4.20 the intersections of $W_{loc}^s(O)$ and $W_{loc}^u(O)$ with Σ_0^s and Σ_0^u are given by $\eta = 0$ and $\eta^* = 0$, respectively. The non-degeneracy condition in these coordinates means that the image of the segment $\eta^* = 0$ on Σ_0^u is transversal at the point $(0,0) \in \Sigma_0^s$ to the segment $\eta = 0$ on Σ_0^s. This is expressed as $(\partial Q_0 / \partial \theta_u^*)(0,0) \neq 0$. Therefore, the existence of a non-degenerate homoclinic orbit γ means that

$$P_0(0,0) = Q_0(0,0) = 0 \quad \text{and} \quad \frac{\partial Q_0}{\partial \theta_u^*}(0,0) \neq 0. \tag{4.21}$$

Up to higher order terms, one can assume that $\Psi_c : V_c^u \to V_c^s$ is given by

$$(\theta_s, \eta) = \Psi_c(\theta_u^*, \eta^*) = \begin{pmatrix} \alpha_c & \beta_c \\ \rho_c & \delta_c \end{pmatrix} \begin{pmatrix} \theta_u^* \\ \eta^* \end{pmatrix} + \begin{pmatrix} A_c \\ B_c \end{pmatrix} \tag{4.22}$$

where from (4.20) and (4.21) it holds

$$\alpha_c \delta_c - \beta_c \rho_c = 1, \quad A_0 = B_0 = 0 \quad \text{and} \quad \rho_0 = \frac{\partial Q_0}{\partial \theta_u^*}(0,0) \neq 0.$$

Note that $\rho_c \neq 0$ if $|c|$ is small enough.

Poincaré return map.   We have to study the dynamics of $\Pi_c = \Psi_c \circ \Phi_c$ in a neighborhood of $\eta = 0$ on the annulus Σ_c^s. Thus, we will consider this map defined on the strip $\mathbb{S}^1 \times [-\varepsilon, \varepsilon]$ where $\varepsilon > 0$ is such that Remark 4.21 can be applied. Notice that Π_c is only well defined on

$$S_c = (\Phi_c)^{-1}(V_c^u) \cap (\mathbb{S}^1 \times [-\varepsilon, \varepsilon])$$

and $\Pi_c(S_c) \subset V_c^s$ but it is not necessarily a subset of S_c. Namely, the first-return map $\Pi_c : S_c \subset \Sigma_c^s \to \Sigma_c^s$ is given by

$$\Pi_c(\theta_s, \eta) = (\beta_c \eta + \alpha_c \theta_s + \alpha_c b_c(\eta) + A_c, \; \delta_c \eta + \rho_c \theta_s + \rho_c b_c(\eta) + B_c). \tag{4.23}$$

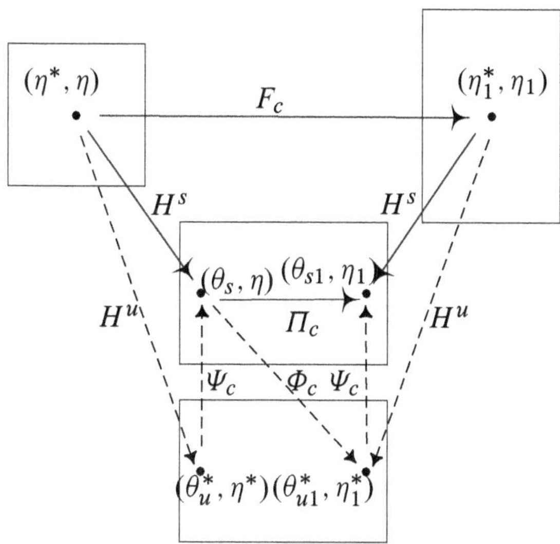

Figure 4.10: *Conjugation*

In fact, we will consider Π_c restricted to the maximal invariant set Δ_c in S_c. That is,

$$\Delta_c = \bigcap_{n \in \mathbb{Z}} \Pi_c^n(S_c).$$

In order to simplify the study of the dynamics of $\Pi_c|_{\Delta_c}$ we will introduce the map

$$F_c : [-\varepsilon, \varepsilon]^2 \to [-\varepsilon, \varepsilon] \times \mathbb{S}^1, \quad F_c(\eta^*, \eta) = (\eta, -\eta^* + \varphi_c(\eta)) \qquad (4.24)$$

where $\varphi_c : [-\varepsilon, \varepsilon] \to \mathbb{R}$ is given by

$$\varphi_c(\eta) = (\alpha_c + \delta_c)\eta + \rho_c b_c(\eta) + \rho_c A_c + (1 - \alpha_c)B_c. \qquad (4.25)$$

Let Ω_c be the maximal invariant set of F_c in $[-\varepsilon, \varepsilon]^2$.

Proposition 4.22. *The restriction $\Pi_c|_{\Delta_c}$ is C^r conjugate to $F_c|_{\Omega_c}$.*

Proof. In Figure 4.10 we show how to define $(\eta_1^*, \eta_1) = F_c(\eta^*, \eta)$. Let us explain this construction in three steps:

1) If $|c|$ is small enough then, for each $(\eta^*, \eta) \in [-\varepsilon, \varepsilon]^2$, with $\varepsilon > 0$ small enough, there exists two unique points $(\theta_s, \eta) = H^s(\eta^*, \eta) \in \Sigma_c^s$ and

$(\theta_u^*, \eta^*) = H^u(\eta^*, \eta) \in \Sigma_c^u$ such that $(\theta_s, \eta) = \Psi_c(\theta_u^*, \eta^*)$. Namely, from (4.22) we get

$$\theta_s = \frac{\alpha_c}{\rho_c}\eta + (\beta_c - \frac{\alpha_c \delta_c}{\rho_c})\eta^* + A_c - \frac{\alpha_c}{\rho_c}B_c \quad \text{and}$$

$$\theta_u^* = \rho_c^{-1}(\eta - \delta_c\,\eta^* - B_c).$$

2) Using the local map Φ_c given in (4.19), the image of $(\theta_s, \eta) \in \Sigma_c^s$ by Φ_c is

$$(\theta_{u1}^*, \eta_1^*) = (\theta_s + b_c(\eta),\ \eta) \in \Sigma_c^u.$$

3) The coordinate η_1 comes from $\Psi_c(\theta_{u1}^*, \eta_1^*) = (\theta_{s1}, \eta_1)$. Namely, substituting in (4.22)

$$\eta_1 = \big(\rho_c\theta_{u1}^* + \delta_c\eta_1^* + B_c\big) = \big(\rho_c\theta_s + \rho_c b_c(\eta) + \delta_c\eta + B_c\big)$$
$$= \big((\rho_c\beta_c - \alpha_c\delta_c)\eta^* + (\alpha_c + \delta_c)\eta + \rho_c b_c(\eta) + \rho_c A_c + (1 - \alpha_c)B_c\big).$$

The coefficient of η^* is $-\det(D\Psi_c)$ and hence

$$(\eta_1^*, \eta_1) = F_c(\eta^*, \eta) = (\eta, -\eta^* + \varphi_c(\eta)), \quad (\eta, \eta^*) \in [-\varepsilon, \varepsilon]^2, \quad (4.26)$$

where φ_c is given in (4.25).

By construction, the map H^s, as defined in the first step, provides a conjugation between $\Pi_c|_{\Delta_c}$ and $F_c|_{\Omega_c}$. $\qquad\qquad\square$

On the other hand, the function

$$\psi_c : [-\varepsilon, \varepsilon] \to \mathbb{R}, \quad \psi_c(\eta) = \rho_c^{-1}\varphi_c(\eta)$$

can be written in the form $u(\eta) + b_c(\eta)$ with

$$u(\eta) = \rho_c^{-1}(\alpha_c + \delta_c)\eta + A_c + \rho_c^{-1}(1 - \alpha_c)B_c.$$

Hence Remark 4.21 is valid for ψ_c. In particular, it follows that φ_c for $|c| > 0$ is a unimodal function with critical point $\eta_c \to 0$ and $\varphi(\eta_c) \to -\infty$ as $c \to 0$.

In what follows, we will study the family (4.26). Since the dynamical behavior of (4.26) for both positive and negatives values of the parameter c is quite similar, for simplicity, we restrict ourselves to the family F_c with parameter $c \geqslant 0$.

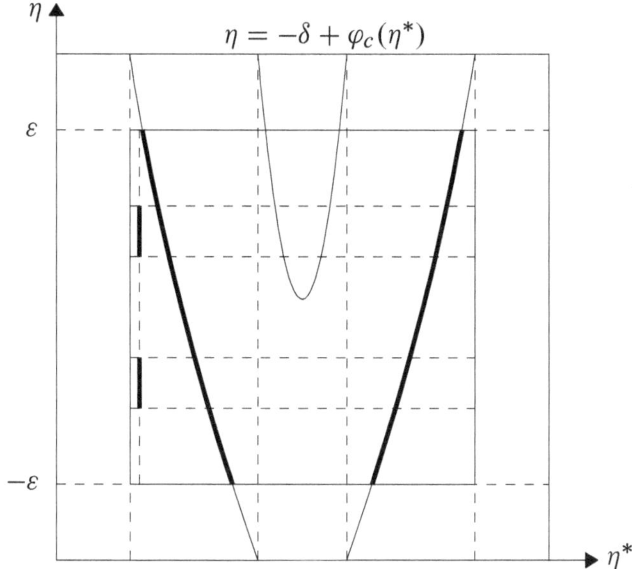

Figure 4.11: *Image by F_c of the vertical segment $\eta^* = \delta$.*

4.4.2 Bidimensional horseshoes

The following theorem shows the existence of a dynamically increasing family of horseshoes (i.e., a continuous family of hyperbolic basic sets conjugate to increasing full shift dynamics) for the one-parametric family of maps F_c. The corresponding version of this theorem for the Poincaré return map Π_c was proved in Lerman (2000, Thm. 1). First we need introduce same espace of symbols. Namely we will consider an alphabet $\{1, 2, \ldots, \pm\infty\}$ consisting of all positive integers and two additional symbols $+\infty$ and $-\infty$. Let Σ_* be the set of bi-sequences in such alphabet satisfying the following rule: only $+\infty$ can follow $+\infty$ and only $-\infty$ can precede $-\infty$.

Theorem 4.23. *For the family F_c given in (4.24), there are $\varepsilon > 0$, $c_0 > 0$ and $\kappa > 0$ such that for every positive $c \leqslant c_0$, the maximal invariant set*

$$\Lambda_c = \bigcap_{n \in \mathbb{Z}} F_c^n(\{(\eta^*, \eta) \in [-\varepsilon, \varepsilon]^2 : |\eta - \eta_c| \geqslant \kappa c^2\})$$

is a hyperbolic set conjugate to the full shift of $n(c)$ symbols where $n(c) \to \infty$ as $c \to 0$. Moreover, the restriction of F_0 to

$$\Omega_0 = \bigcap_{n \in \mathbb{Z}} F_0^n([-\varepsilon, \varepsilon]^2)$$

is conjugate to the Bernoulli shift map $\tau : \Sigma_* \to \Sigma_*$.

The proof of this result follows from standard arguments to construct horseshoes as illustrated in Figure 4.11. More details in Barrientos, Ibáñez, and Rodríguez (2016).

A bi-sequence of the type $(\ldots, -\infty, -\infty, \xi_1, \ldots, \xi_n, \infty, \infty, \ldots)$ corresponds to a homoclinic orbit of a point p that emerges from $\eta^* = 0$ on Σ_0^u (the trace of $W_{loc}^u(O)$) and then flows through a neighborhood of γ intersecting n times the section Σ_0^s before reaching $\eta = 0$ (the trace of $W_{loc}^s(O)$). We refer to such a homoclinic orbit as a n-pulse homoclinic orbit (see Definition 2.10). Consequently the following corollary also holds:

Corollary 4.24 (Belyakov (1984b) and Lerman (2000)). *Under the assumption **(BH)**, **(B1)** and **(B2)**, for each tubular neighborhood $\mathcal{T}$ of Γ and $n \in \mathbb{N}$ there exist non-degenerate n-pulse bifocus homoclinic orbits in $\mathcal{T}$.*

4.4.3   Nearby heterodimensional cycles and tangencies

The goal of this section is to explain how C^1 robust heterodimensional cycles and C^1 robust homoclinic tangencies could be obtained by arbitrarily small perturbation of first return map of (4.1) under the assumptions **(BH)**, **(B1)** and **(B2)**. We will define properly below these objects, but we address the reader to Bonatti, Díaz, and Viana (2005) to see the dynamical consequences of such configurations. As usual, we will use the term *suspended* in order to empathized that cycles and tangencies will be obtained for the Poincaré first-return map. First of all we recall the formal definition of these concepts.

A diffeomorphism f has a *homoclinic tangency* associated with a transitive hyperbolic set Λ if there is a pair of points $x, y \in \Lambda$ such that the stable manifold $W^s(x)$ of x and the unstable manifold $W^u(y)$ of y have some non-transverse intersection. The tangency is said to be C^1 *robust* if there is a C^1 neighborhood $\mathcal{U}$ of f such that, for every diffeomorphism $g \in \mathcal{U}$, the continuation Λ_g of Λ has a homoclinic tangency.

A diffeomorphism f has a *heterodimensional cycle* if there exist two transitive hyperbolic sets Λ and Γ with different stability indices such that

$$W^s(\Lambda) \cap W^u(\Gamma) \neq \emptyset \quad \text{and} \quad W^u(\Lambda) \cap W^s(\Gamma) \neq \emptyset.$$

This cycle is said to be C^1 *robust* if there is a C^1 neighborhood $\mathcal{U}$ of f such that, for every diffeomorphism $g \in \mathcal{U}$, there is a heterodimensional cycle associated with the continuations Λ_g and Γ_g of Λ and Γ respectively.

Theorem 4.25. *Under the assumptions **(BH)**, **(B1)** and **(B2)**, every C^1 neighborhood of X contains vector fields with both, (suspended) C^1 robust heterodimensional cycles and (suspended) C^1 robust homoclinic tangencies.*

This theorem is a consequence of a useful criterium to yield C^1 robust heterodimensional cycles and homoclinic tangencies that we explain below. To do this, we need first introduce some additional concepts of diffeomorphisms.

Let f be a diffeomorphism with a n-periodic point p. Assume that there is a Df-invariant partially hyperbolic splitting $E^{ss} \oplus E^c \oplus E^{uu}$ defined over the orbit $\mathcal{O}(p)$ of p such that E^c has dimension one, every eigenvalue λ of $Df^n(p)$ corresponding to E^{ss} satisfies $|\lambda| < 1$, and every eigenvalue β of $Df^n(p)$ corresponding to E^{uu} satisfies $|\beta| > 1$. The partial hyperbolicity means that if λ_c is the eigenvalue of $Df^n(p)$ corresponding to E^c, then $|\lambda| < |\lambda_c| < |\beta|$ for every pair of eigenvalues λ and β corresponding to E^{ss} and E^{uu}, respectively. If the periodic point p is not hyperbolic, either $\lambda_c = 1$ or $\lambda_c = -1$. We say that p is a *saddle-node* in the first case and a *flip* in the second one. The *strong stable manifold* $W^{ss}(p)$ of p for f is the unique f^n-invariant manifold which is tangent to E_p^{ss} and satisfies $\dim W^{ss}(p) = \dim E_p^{ss}$. The *strong unstable manifold* $W^{uu}(p)$ of p for f is defined in a similar way, but considering the bundle E_p^{uu}. The periodic point p has a *strong homoclinic intersection* if there is

$$x \in W^{ss}(p) \cap W^{uu}(p) \qquad \text{with } x \neq p.$$

We refer to x as a strong homoclinic point. A strong homoclinic intersection or point is said *quasi-transverse* if

$$T_x W^{ss}(p) + T_x W^{uu}(p) = T_x W^{ss}(p) \oplus T_x W^{uu}(p).$$

Otherwise, it is said *tangential*.

The following criterion was obtained in Barrientos, Ibáñez, and Rodríguez (2016) as consequence of previous results due to Bonatti and Díaz (Bonatti and Díaz (2008, Theorem 2.4), Bonatti and Díaz (2012, Theorem 4.9) (see also Bonatti and Díaz (2012, Proposition 5.4) and Barrientos, Ki, and Raibekas (2014, Theorem A)).

Theorem 4.26. *Let f be a diffeomorphism with both, quasi-transverse and tangential strong homoclinic intersections associated with a saddle-node periodic point. Then there are diffeomorphisms arbitrarily C^1 close to f with both, a C^1 robust heterodimensional cycle and a C^1 robust homoclinic tangency.*

Now, we will apply this criterion to get Theorem 4.25. We recall that the Poincaré return map Π defined on a transversal section, the solid torus Σ^s, of a non-degenerated bifocus homoclinic orbit γ of a Hamiltonian vector field X can be written, in appropriate coordinates, as

$$\Pi(\theta^s, \eta, c) = (\Pi_c(\theta^s, \eta), c), \qquad \theta^s \in \mathbb{S}^1, \ |\eta| \leq \varepsilon, \ |c| \leq c_0$$

for $\varepsilon, c_0 > 0$ small enough. Here Π_c is the symplectic map on the annulus $\Sigma_c^s = \Sigma^s \cap H^{-1}(c)$ given in (4.23). In fact, this map Π_c is only well defined, for $\varepsilon > 0$ small enough, on the domain

$$S_c = (\Phi_c)^{-1}(V_c^u) \cap (\mathbb{S}^1 \times, [-\varepsilon, \varepsilon])$$

with $V_c^u = V^u \cap \Sigma^u \cap H^{-1}(c)$ for a neighborhood V^u of the transversal intersection q^u between the homoclinic orbit γ and a solid torus Σ^u.

Set $I = [-c_0, c_0]$. In Proposition 4.22, for each $c \in I$, we have shown that Π_c restricted to the maximal invariant set Δ_c in S_c is C^r conjugate to the restriction of F_c given in (4.24) to its maximal invariant set Ω_c. Therefore, it follows that Π restricted to

$$\Delta_c \rtimes I \stackrel{\text{def}}{=} \{(z, c) : c \in I \text{ and } z \in \Delta_c\}$$

is C^1 conjugate to

$$f : \Omega_c \rtimes I \to \Omega_c \rtimes I, \quad f(z, c) = (F_c(z), c)$$

where $\Omega_c \rtimes I \stackrel{\text{def}}{=} \{(z, c) : c \in I \text{ and } z \in \Omega_c\}$. According to Theorem 4.23, for each $c \in I, c \neq 0$, there is a hyperbolic basic set Λ_c of F_c in $[-\varepsilon, \varepsilon]^2$ such that $F|_{\Lambda_c}$ is conjugate to the Bernoulli shift of $n(c) \geq 2$ symbols. Moreover, since the number of symbols $n(c)$ associated with the families of horseshoes Λ_c goes to infinite as $c \to 0$ necessarily the creation and destruction of these horseshoes is by means of homoclinic tangencies. We fix c at one of these bifurcation values for which F_c has a non-transversal homoclinic intersection associated with a saddle fixed point $p_c \in \Lambda_c$. Hence, $P = (p_c, c)$ is a saddle-node fixed point of f with a tangential strong homoclinic intersection. Moreover, since Λ_c is a horseshoe of F_c, $W^s(p_c)$ meets transversally $W^u(p_c)$ in a point $q_c \neq p_c$ and hence,

$$Q = (q_c, c) \in W^{ss}(P) \cap W^{uu}(P) \quad \text{with} \quad Q \neq P.$$

Thus, f has both, a quasi-transverse and a tangential strong homoclinic intersection. Applying Theorem 4.26 we complete the proof of Theorem 4.25.

4.5 Reversible bifocus homoclinic orbits

Here we will give a rigorous description of the dynamics around of a reversible (not necessarily conservative) non-degenerate bifocus homoclinic orbit. Specifically, we will assume that (4.1) satisfies the following property:

(BR) The vector field X is R-reversible. That is, there a linear map $R : \mathbb{R}^4 \to \mathbb{R}^4$ such that

$$R^2 = \text{id} \quad \text{and} \quad X \circ R = -R \circ X.$$

As a consequence of **(BR)**, if $u(t)$ is a solution of (4.1), then so is $Ru(t)$. We say that a trajectory of (4.1) is *reversible* or *symmetric* if it is invariant under the involution R. We also assume that O is a bifocus equilibrium as in **(B1)**. Moreover, we suppose that we have at least one non-degenerated homoclinic orbit γ as in **(B2)** but with the extra assumption that this orbit is symmetric:

(B2') There is a non-degenerate homoclinic orbit γ to O such that $R(\gamma) = \gamma$.

Observe that the reversibility of the homoclinic orbit in **(B2')** implies that $R(O) = O$. Consequently, the derivative $DX(O)$ is also R-reversible. Indeed, since $R \circ X = -X \circ R$, differentiating both sides of the equality and using that R is linear and $R(O) = O$, it follows that $R \circ DX(O) = -DX(O) \circ R$. This reversibility of the linear part of X at O bring as a consequence that

$$\dim \mathrm{Fix}(R) = 2 \quad \text{where} \ \ \mathrm{Fix}(R) = \{u : R(u) = u\} \tag{4.27}$$

and the eigenvalues of $DX(O)$ are exactly of the form

$$-\alpha \pm i\omega \quad \text{and} \quad \alpha \pm i\omega \quad \text{where } \alpha > 0 \text{ and } \omega \neq 0. \tag{4.28}$$

In particular we are under the assumption **(B4)**. To see (4.27) and (4.28), without loss of generality we can take $\mathbb{R}^4 = E \oplus F$ where $F = \mathrm{Fix}(R)$ and $E = \{u : R(u) = -u\}$. Since $DX(O)$ is R-reversible we have that $DX(O)F \subset E$ and $DX(O)E \subset F$. Hence, since $DX(O)$ is a hyperbolic linear matrix, $\dim F = \dim E$ and therefore $\dim \mathrm{Fix}(R)$ must to be equal two. Finally, (4.28) follows from the fact that, by the R-reversibility of $DX(O)$,

$$\det[DX(O) - \lambda I] = \det[DX(O) + \lambda I]$$

and then if λ is a zero of the characteristic polynomial, then $-\lambda$ is also a zero of the same polynomial.

The cycle $\Gamma = \{O\} \cup \gamma$ under the assumptions **(BR)**, **(B1)** and **(B2')** is usually referred as (non-degenerate) *reversible bifocus homoclinic cycle*. The similarity between reversible and Hamiltonian systems has been demonstrated in many cases. For instance, both reversible and Hamiltonian homoclinic orbits are accompanied by a one-parameter family of periodic orbits Devaney (1976a), Devaney (1977), and Vanderbauwhede and Fiedler (1992). Also, as we have shown in Corollary 4.24 that in any tubular neighborhood of a non-degenerate Hamiltonian bifocus homoclinic orbit there are infinitely many secondary homoclinic orbits. That is, other bifocus homoclinic orbits which make several excursions along the primary homoclinic orbit. These results were extended for non-degenerate reversible bifocus homoclinic cycles by Härterich (1998) showing that the secondary homoclinics are also reversible and non-degenerate.

Theorem 4.27 (Härterich (1998)). *Under the assumptions **(BR)**, **(B1)** and **(B2')**, in any tubular neighborhood of Γ, there exist infinitely many reversible non-degenerate homoclinic orbits to O. Moreover, each homoclinic orbit is accumulated by a one-parameter family of reversible periodic orbits.*

We fix a tubular neighborhood $\mathcal{T}$ of Γ and $N \geqslant 1$. According to Theorem 4.27 we can find N different reversible non-degenerate homoclinic orbits γ_i in $\mathcal{T}$. Observe that for $N = 1$ we are only claimed the presence of a unique homoclinic orbit and Theorem 4.27 is not used. Consider the network

$$\Gamma_N = \{O\} \cup \gamma_1 \cup \cdots \cup \gamma_N.$$

We will perform a similar analysis to that of Barrientos, Raibekas, and Rodrigues (2019), Härterich (1998), and Ibáñez and Rodrigues (2015) and study the first return map over a cross-section transverse to the homoclinic network Γ_N.

4.5.1   The first-return map

Without restriction, we can assume that the linear involution R in **(BR)** is given by $R(x_1, x_2, y_1, y_2) = (y_1, y_2, x_1, x_2)$. Thus, using bipolar coordinates $(r_s, \theta_s, r_u, \theta_u)$, the two-dimensional set of fixed points by R is written as

$$\mathrm{Fix}(R) = \{r_s = r_u, \theta_s = \theta_u\}.$$

Analogously to §4.2, we consider the three-dimensional cross-sections near the origin, Σ^s and Σ^u, which are solid tori defined in (4.3). Observe that we have $\Sigma^u = R(\Sigma^s)$. Moreover, we choose this section small enough so that

$$\{q_i^s\} = \gamma_i \cap \Sigma^s \subset W_{loc}^s(O) \quad \text{for all } i = 1, \ldots, N.$$

By reversibility we also have that $q_i^u = R(q_i^s)$ so that

$$\{q_i^u\} = \gamma_i \cap \Sigma^u \subset W_{loc}^u(O) \quad \text{for all } i = 1, \ldots, N.$$

As in (4.4), we get a local map $\Phi : \Sigma^s \setminus W^s(O) \to \Sigma^u$ defined by

$$r_s^* = r_u \qquad \theta_s^* = \theta_s - \frac{\omega}{\alpha} \ln r_u \qquad \theta_u^* = \theta_u - \frac{\omega}{\alpha} \ln r_u. \tag{4.29}$$

We can make use of the reversibility in the construction of the global Poincaré map by taking a R-invariant cross-section Σ containing the points

$$\{q_i\} = \gamma_i \cap \mathrm{Fix}(R) \quad \text{for } i = 1, \ldots, N.$$

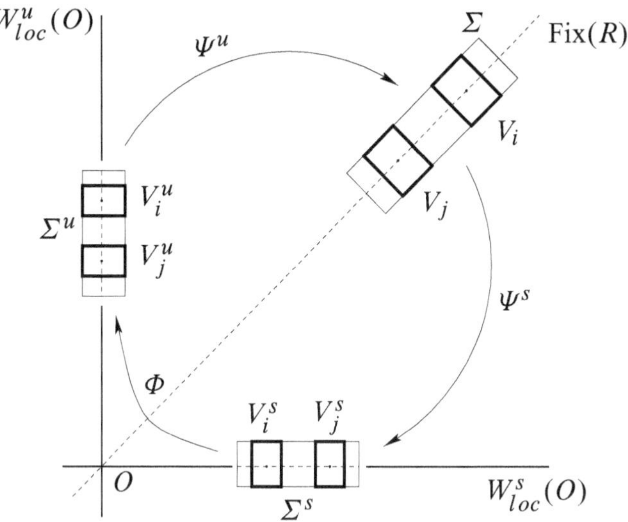

Figure 4.12: *First return map $\Pi = \Psi^u \circ \Phi \circ \Psi^s$ on the section Σ.*

A global map Ψ^u between Σ^u and Σ is induced by the flow along γ_i. To be more specific, given any neighborhood V_i of q_i in Σ, there exists a neighborhood $V_i^u \subset \Sigma^u$ of q_i^u and a map $\tau_i : V_i^u \to \mathbb{R}$ such that

$$\varphi(\tau_i(q_i^u), q_i^u) = q_i \quad \text{and} \quad \Psi_i^u(x) \overset{\text{def}}{=} \varphi(\tau_i(x), x) \in V_i$$

for all $x \in V_i^u$ and $i = 1, \ldots, N$. By taking the sets $V_i \subset \Sigma$ small enough we can obtain that V_i^u are pairwise disjoint compact neighborhoods of q_i^u in Σ^u for all $i = 1, \ldots, N$. Moreover, we can also take V_i such that $R(V_i) = V_i$. Hence, Ψ^u is defined as

$$\Psi^u|_{V_i^u} = \Psi_i^u \quad \text{for all } i = 1, \ldots, N.$$

By reversibility, Ψ^s is the semi-global map between Σ and Σ^s given by

$$\Psi^s = R \circ (\Psi^u)^{-1} \circ R.$$

Finally, we introduce the Poincaré first return map on Σ following the homoclinic network Γ_N as $\Pi = \Psi^u \circ \Phi \circ \Psi^s$ (see Figure 4.12). Observe that Π is actually defined as the composition of three maps: first Ψ^s which is well defined from $V = R(V) \subset \Sigma$ to Σ^s, then $\Phi : \Sigma^s \setminus W^s_{loc}(O) \to \Sigma^u$ and finally $\Psi^u : V^u \to \Sigma$ where

$$V = V_1 \cup \cdots \cup V_N \subset \Sigma \quad \text{and} \quad V^u = V_1^u \cup \cdots \cup V_N^u \subset \Sigma^u.$$

On the other hand,

$$V_i^s \stackrel{\text{def}}{=} \Psi^s \circ R(V_i) = R(V_i^u)$$

is a compact neighborhood of q_i^s in Σ^s for $i = 1, \ldots, N$. Moreover, we also have that Π is a reversible map, i.e., $R \circ \Pi \circ R = \Pi^{-1}$.

4.5.2   Switching

We take discs

$$D_i \subset V_i \cap \text{Fix}(R) \subset \Sigma \text{ centered at } q_i \text{ for all } i = 1, \ldots, N.$$

Now, we introduce the local stable and unstable manifolds of O in Σ as

$$W_i^u = \Psi^u\left(W_{loc}^u(O) \cap V_i^u\right) \quad \text{and} \quad W_i^s = (\Psi^s)^{-1}\left(W_{loc}^s(O) \cap V_i^s\right)$$

for $i = 1, \ldots, N$. Observe that the orbits starting in W_i^s (resp. W_i^u) goes directly to O, in forward (resp. backward) time. We will need the following basic result.

Lemma 4.28. *If $x \in W_i^s \cap \Pi^n(\text{Fix}(R))$ for some $n \geq 0$ then the associated solution is a reversible homoclinic orbit. Similarly, if $x \in \text{Fix}(R) \cap \Pi^n(\text{Fix}(R))$ for $n \geq 1$ then the associated solution is a reversible periodic orbit.*

Proof. We prove that $x \in W_i^s \cap \Pi^n(\text{Fix}(R))$. We have that $\Pi^{-n}(x) \in \text{Fix}(R)$ and hence, by the reversibility, $\Pi^{-2n}(x) \in W_i^u$. Thus the orbit associated with x is a homoclinic orbit. Similarly, if $x \in \text{Fix}(R) \cap \Pi^n(\text{Fix}(R))$ then $\Pi^n(x) = R \circ \Pi^{-n} \circ R(x) = R \circ \Pi^{-n}(x) = \Pi^{-n}(x)$. Therefore, $\Pi^{2n}(x) = x$. $\qquad\square$

If $\mathcal{R}$ is a measurable set of $\text{Fix}(R) \cap \mathcal{T}$ let us denote by $\mathcal{A}(\mathcal{R})$ its usual area.

Proposition 4.29. *For any $k \geq 1$, $n \in \mathbb{Z}^{k-1}$ and $i \in \{1, \ldots, N\}^k$ there exist pairwise disjoints compact sets diffeomorphic to a disc*

$$D_{ij}^{nm} \subset \text{Fix}(R) \quad \text{for all } m \in \mathbb{Z} \text{ and } j = 1, \ldots, N$$

such that

1. *$D_{ij}^{nm} \subset D_i^n$,*

2. *$\Pi^k(D_{ij}^{nm}) \subset V_j$,*

3. *there is $q_{ij}^{nm} \in D_{ij}^{nm}$ such that $\Pi^k(q_{ij}^{nm}) \in W_j^s$ and*

4. *there is $\lambda < 1$ such that $\mathcal{A}(D_{ij}^{nm}) < \lambda \cdot \mathcal{A}(D_i^n)$.*

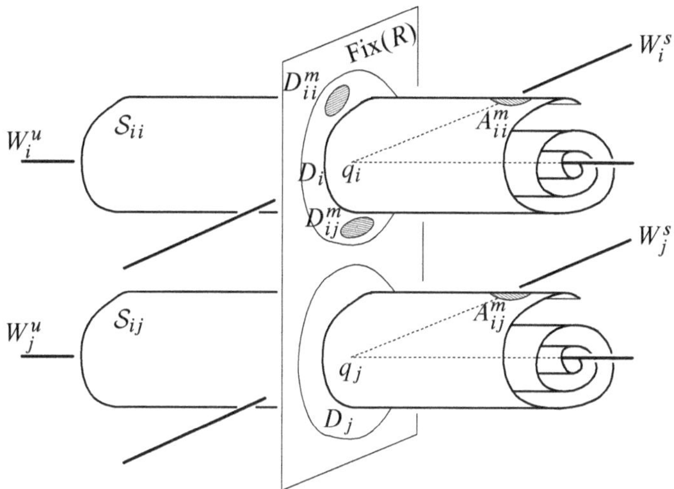

Figure 4.13: *Spiralling geometry for the return map on section Σ.*

This result follows by induction applying Proposition 4.3 and Lemma 4.28 as illustrated in Figure 4.13. See Barrientos, Raibekas, and Rodrigues (2019) for more details.

Let

$$H = \{q_i^n : i \in \{1, \ldots, N\}^k, n \in \mathbb{Z}^{k-1}, k \in \mathbb{N}\}.$$

According to Lemma 4.28 and Lemma 4.29 the flow orbit associated with any point $q \in H$ is a reversible non-degenerate homoclinic orbit. In particular, for $N = 1$, this observation proves the existence of infinitely many non-degenerate bifocus homoclinic orbits in any tubular neighborhood of $\Gamma = \{O\} \cup \gamma$ as Theorem 4.27 asserted. Moreover, one also can find a one-parameter family of reversible periodic orbits accumulating on q. Indeed, by construction we can find, $k \geqslant 1$ and a small neighborhood D of q in Fix(R) such that $\Psi^s \circ \Pi^{k-1}(D)$ is diffeomorphic to a two-dimensional disc transverse to $W_{loc}^s(O)$. Hence, $\Pi^k(D \setminus \{q\})$ contains a spiraling sheet which intersect transversally Fix(R) into a spiral. By Lemma 4.28, this curve is formed by initial condition of periodic orbits. In particular, this provides a one-parameter family of periodic orbits accumulating on q as indicated in Theorem 4.27.

Now, consider

$$K = \bigcap_{k \in \mathbb{N}} \bigcup_{n \in \mathbb{Z}^{k-1}} \bigcup_{i \in \{1,\ldots,N\}^k} D_i^n. \tag{4.30}$$

From Proposition 4.29 it follows that K is an non-empty set and $K \subset \overline{H}$. Moreover, for any pair of sequences $i = (i_k)_{k \in \mathbb{N}} \in \Sigma_N^+ = \{1, \ldots, N\}^{\mathbb{N}}$ and $n = (n_k)_{k \in \mathbb{N}} \in \mathbb{Z}^{\mathbb{N}}$, there is a unique point $x_i^n \in K$ such that

$$\{x_i^n\} = \bigcap_{k \in \mathbb{N}} D_{i_1 \ldots i_k}^{n_1 \ldots n_{k-1}}. \tag{4.31}$$

In fact, we have the following:

Remark 4.30. *Any point in K is uniquely identified by a pair $(i, n) \in \Sigma_N^+ \times \mathbb{Z}^{\mathbb{N}}$.*

On the other hand, Proposition 4.29 and Equation (4.31) imply that

$$\Pi^k(x_i^n) \in V_{i_{k+1}} \quad \text{for all } k \geqslant 0.$$

Since $x_i^n \in \text{Fix}(R)$, then by the reversibility we also have that $\Pi^{-k}(x_i^n) \in V_{i_{k+1}}$ for all $k \geqslant 1$. Recalling the notion of *switching* in Definition 2.23, we have proved the following result:

Proposition 4.31. *The network Γ_N exhibits switching by reversible trajectories. Moreover, the starting point of the orbit realization can be taken in K.*

But dynamics around the network Γ_N is still more complicated. A trajectory $\ll$ is said to be a *super-homoclinic orbit* to Γ_N if $\ll$ is a bi-asymptotic connection to Γ_N, that is it accumulates on the network Γ_N in forward and backward time. We say that the homoclinic network Γ_N exhibits *symmetric super-homoclinic switching* if any tubular neighborhood $\mathcal{T}$ of Γ_N and for each $\omega \in \Sigma_N^+$ we find a super-homoclinic orbit in $\mathcal{T}$ realizing ω with starting point $x_\omega \in \text{Fix}(R)$. Observe that since x_ω belongs to $\text{Fix}(R)$, then the super-homoclinic orbit also follows the sequence ω in backward time. Similarly, when any prescribed finite path is realized by a reversible homoclinic (resp. periodic) orbit starting in $\text{Fix}(R)$, we say that Γ_N exhibits *symmetric homoclinic* (resp. *periodic*) *switching*.

Theorem 4.32 (Barrientos, Raibekas, and Rodrigues (2019)). *Under the assumption (BR), (B1) and (B2'), for any tubular neighborhood $\mathcal{T}$ of Γ and for $N \geqslant 2$, there are different non-degenerate reversible bifocus homoclinic orbits $\gamma_1, \ldots, \gamma_N$ in $\mathcal{T}$ such that the homoclinic network $\Gamma_N = \{O\} \cup \gamma_1 \cup \cdots \cup \gamma_N$ exhibits symmetric super-homoclinic, homoclinic and periodic switching.*

As we have showed in §4.3.1 and §4.4.2, in the presence of a non-degenerate Shilnikov or conservative bifocus homoclinic cycle we can find suspended horseshoes on any number of symbols arbitrarily close to the connection. However, nothing is known in general about the presence of horseshoes in a neighborhood of a reversible bifocus homoclinic cycle $\Gamma = \{0\} \cup \gamma$. Homburg and Lamb studied in Homburg and Lamb (2006) this situation under the extra assumption that there is an orbit γ_{a_0} in the one-parameter family γ_a of accompanying periodic orbits to γ whose stable and unstable manifolds intersect in a reversible bi-asymptotic orbit ρ_{a_0} to γ_{a_0}. They concluded that the non-wandering set of the return map describing the dynamics near ρ_{a_0} is contained in a set with a lamination of one-dimensional leaves parameterized by a subshift of finite type, that is similar as in the Hamiltonian case. On the other hand, in the general case, observe that Theorem 4.32 also concludes, in particular, the presence of sets of initial conditions arbitrarily close to the cycle Γ whose dynamics is semi-conjugate to the unilateral shift on N-symbols for any $N \geq 2$ and have a dense set of periodic orbits. This could be seen as a weak result on the presence of chaotic dynamics near of a non-degenerate reversible bifocus homoclinic orbit. See Barrientos, Raibekas, and Rodrigues (2019).

4.5.3 Nearby heterodimensional cycles

The goal of this section is to explain the following result on the creation of heterodimensional cycles by arbitrarily small perturbation in the reversible case.

Theorem 4.33. *Every vector field X under the assumption **(BR)**, **(B1)** and **(B2')** can be C^1 approximated by vector fields exhibiting suspended C^1 robust heterodimensional cycles.*

Proof. In order to prove the theorem, we consider the intersection point $q_i \in \Sigma$ of the non-degenerate homoclinic orbit γ_i with $\mathrm{Fix}(R)$. As we have shown, this point is accumulated by a one-parameter family $(p_a)_a$ of reversible periodic orbits p_a of Π. It follows from (4.8) that $\Phi^s(p_a)$ have two clearly hyperbolic directions for the return map $\Psi^s \circ \Pi \circ (\Psi^s)^{-1}$. Namely, there are enormous stretching in the θ_u-direction and contraction in the θ_s-direction. This implies that if p_a is close enough to q_i the strong stable and strong unstable local manifolds $W^{ss}_{loc}(p_a)$ and $W^{uu}_{loc}(p_a)$ of p_a for Π are one-dimensional manifolds parallel to W^s_i and W^u_i respectively. On the other hand, as a consequence of the reversibility, we are under the assumption **(B4)** (i.e., $\delta = 1$) and thus the radial direction is neutral. This direction corresponds exactly with spiral family of periodic orbits. Thus, p_a are saddle-node periodic points of Π. Now we will prove that by an arbitrarily small perturbation we can create a strong homoclinic intersection associated with some periodic point p_a close enough to q_i. From Theorem 4.26 (c.f. Bonatti and Díaz (2012, Thm. 2.4)) one gets a diffeomorphism C^1 arbitrarily close to Π with a C^1 robust heterodimensional cycle concluding the proof of the theorem.

Then, in order to provide the strong homoclinic intersection, observe that the image by (4.29) of a small segment in Σ^s parallel to the $W^u(O)$ is a helix surround $W^u_{loc}(O) \cap \Sigma^u$. In fact, if we assume that the small segment joints $r_u = b_{k+1}$ with $r_u = b_k$ then its image joints $r_s^* = a_{k+1}$ with $r_s^* = a_k$ where a_k and b_k are given in (4.6) and (4.7) respectively. Thus, since $W^{uu}_{loc}(p_a)$ can be taken arbitrarily close to W^u_i, we can see $\Psi^s(L)$ as a small segment in Σ^s as above (joining $r_u = b_{k+1}$ with $r_u = b_k$ for k large) where L is a compact disc in $W^{ss}_{loc}(p_a) \setminus \{p_a\}$. Hence $\hat{L} = \Phi \circ \Psi^s(L)$ is an helix surround $W^u_{loc}(O) \cap \Sigma^u$ joining $r_s^* = a_{k+1}$ with $r_s^* = a_k$. Thus $\Pi(L) = \Psi^u(\hat{L})$ contains a piece L_i of a helix contained in a cylinder centered at W^s_i with inner radius a_{k+1} and outer radius a_k. In fact, since $\delta = 1$ we have $a_k = b_k$ and thus we can say that the order of the proximity of $W^{ss}_{loc}(p_a)$ and L_i to q_i is the same. By means of a small perturbation of Π supported on a neighborhood of L we can connect $W^{ss}_{loc}(p_a)$ and $L_i \subset \Pi(L)$. From the non-degenerate condition of γ_i this intersection is quasi-transverse and the perturbation does not affect the saddle-node p_a. Therefore, we get a strong homoclinic intersection associated with a saddle-node periodic point as required and complete the proof. $\qquad\square$

Another constructive proof of the above result can be explained as follows. Since W^u_i and W^s_i are quasi-transverse and the unstable manifold of the one-parametric family $(p_a)_a$ of periodic orbits p_a surrounded q_i is a spiraling sheet accumulating on W^u_i we can find a small disc transversally intersecting W^s_i. Hence, the image of this disc by Π is a spiraling sheet accumulating on W^u_j for $j \neq i$. Observe that the point $\{q_j\} = \Sigma \cap \gamma_j$ is also accumulated by a one-parameter family $(\tilde{p}_a)_a$ of the periodic orbits $\tilde{p}_a$. Moreover, the stable manifold of this one-parameter family is a spiraling sheet accumulating on W^s_j. In particular, from the non-degenerate condition (quasi-transversality between W^u_j and W^s_j) we have a transversal intersection between these spiraling sheets, i.e., between the unstable manifold of the family $(p_a)_a$ and the stable manifold of the family $(\tilde{p}_a)_a$. On the other hand, by the same argument as in the proof of the above theorem one shows that we can find a strong stable local manifold $W^{ss}_{loc}(p)$ of a periodic point $p = p_a$ arbitrarily close of the strong unstable manifold $W^{uu}(\tilde{p})$ of a periodic point $\tilde{p} = \tilde{p}_a$. Then by a small perturbation we can connect both. In fact, perturbing the saddle-nodes p and $\tilde{p}$ we can assume that both are hyperbolic with different indices so that the transversal spiral sheets corresponding to the unstable and stable manifolds of $(p_a)_a$ and $(\tilde{p}_a)_a$ are now the unstable and stable manifolds of p and $\tilde{p}$ respectively. Then we obtain a heterodimensional cycle (of co-index one) associated with $\tilde{p}$ and $\tilde{q}$. The C^1 robustness can be obtained now from Bonatti and Díaz (2008, Thm. 1.5).

An interesting final remark is the following:

Remark 4.34. *The heterodimensional cycles obtained in the above theorem (even in Theorem 4.25) are obtained without breaking the homoclinic connection. Thus one gets vector fields satisfying **(B1)** and **(B2)** having suspended heterodimensional cycles. Of course, these vector fields are neither Hamiltonian nor reversible.*

The existence of heterodimensional cycles near the Shilnikov bifocus homoclinic cycles (where there is no resonance of the eigenvalues) is an open problem as well as its existence for generic unfoldings of nilpotent singularities where these kinds of cycles are present. See Chapter 5.

5   *Singularities and chaos*

In this chapter we pay attention to the study of singularities of vector fields and their unfoldings. Our main goal is to provide a catalogue of singularities for which results are available that guarantee the genesis of chaotic dynamics in their neighborhood, but we also provide a very general overview about the state of the art regarding the unfolding of certain singularities. We must emphasize again that chaotic behaviors are not easy to detect in given models, but singularities are, definitely, much more manageable objects.

It is also remarkable that, besides a purely academic interest and the unquestionable benefits of having elementary criteria to prove the existence of chaos, a huge number of real-world applications lead to models consisting of families of vector fields, the main setting along this whole chapter.

If we are going to deal with local bifurcations of low codimension, it appears natural to recall first the results about classification of singularities and also the analysis of some of the most elementary bifurcations. Thus, with simple examples, reader familiarizes with some of the basic concepts in local bifurcation theory: singularity, unfolding, codimension,…Later, we focus on nilpotent singularities, that is, singularities for which the linear part does not vanish but all eigenvalues are zero. The unfolding of the two-dimensional nilpotent singularity of codimension two is nothing more than the well-known Bogdanov–Takens bifurcation. In spite of being a classic bifurcation diagram, we include a brief discussion focused on the appearance of a homoclinic bifurcation curve. This provides a good illustration for some of the ideas required in higher dimensional cases. Nilpotent singularities with dimensions three and four are the core of this chapter. In §5.6, we see how the three-dimensional nilpotent singularity of codimension three unfolds, generically, Bykov cycles which, as showed in Chapter 3, give rise Shilnikov homoclinic orbits, and hence, as explained in Chapter 2, also persistent non-hyperbolic strange attractors. §5.7 deals with the unfolding of the four-dimensional nilpotent singularity of codimension four. We show how bifocus homoclinic cycles are exhibited. The emergence of chaos in a neighborhood of these cycles is discussed in Chapter 4. To conclude, we deal with other singularities which also imply the existence of strange attractors, mainly with Hopf–Zero singularities. Nevertheless, because the required techniques to show the existence of chaos are quite different to those employed with nilpotent cases, a complete discussion has been discarded.

5.1   Classification of singularities

Given a vector field X, any point where X equals 0 is said to be a singularity. In this respect, singularity and equilibrium are the same thing, but we often refer as singularities to vector fields themselves, when they exhibit an equilibrium point with specific properties.

The first step to study singularities is their classification. To do this we need to introduce a topology and an equivalence relation with dynamical meaning. Notions of germ, k-jet equivalence and C^0 equivalence come into play.

Consider C^∞ vector fields X and Y on $\mathbb{R}^n$ with $X(0) = Y(0) = 0$. We say that X and Y are germ-equivalent if there exists a neighborhood U of 0 such that $X\mid_U = Y\mid_U$. This relation is obviously an equivalence and the corresponding classes are called *germs*. Let G^n be the set of all germs in 0 of C^∞ vector fields.

Given $X, Y \in G^n$, they are said to be k-jet equivalent, with $k \in \mathbb{N} \cup \infty$, if their Taylor expansions at 0 up to order k are equal. It is again an equivalence relation and we call k-jets to the corresponding classes. Given $X \in G^n$, its class of equivalence is said to be the k-jet of X and denoted by $j_k X$. The set of k-jets in G^n is denoted by J^n_k. It should be noticed that there exists a one-to-one correspondence between J^n_k and the space of vector fields on $\mathbb{R}^n$ exhibiting a singularity at 0 and whose components are polynomials of degree less than or equal to k. Therefore we can consider J^n_k endowed with the Euclidean topology. Let j_k be the projection map of G^n on the quotient set J^n_k. We consider G^n endowed with the coarsest topology which makes continues j_k for all $k \in \mathbb{N}$.

A set $A \subset G^n$ is said to be analytic (resp. algebraic) if there exists $k \in \mathbb{N}$ and an analytic (resp. algebraic) set $\widetilde{A} \subset J^n_k$ such that $A = j_k^{-1}(\widetilde{A})$. The codimension of A is defined as the codimension of $\widetilde{A}$ as subset of a finite dimensional space. Regular manifolds in G^n are similarly defined.

With regard to a dynamical classification, given $X, Y \in G^n$, they are said C^0 equivalent if for some representatives $\widetilde{X}$ and $\widetilde{Y}$ of X and Y, respectively, there exist open neighbourhoods U and V of $0 \in \mathbb{R}^n$ and a homeomorphism $h : U \to V$ such that h sends orbits of $\widetilde{X}$ to orbits of $\widetilde{Y}$ preserving the sense, but not necessarily the parametrization. If parametrization is preserved, we say that X and Y are C^0 conjugated. Notions of C^0 equivalence and C^0 conjugacy permit to identify behaviors which are qualitatively the same.

The topological classification of singularities hangs on being able to stratify the space of germs as

$$G^n = V_0^n \supset V_1^n \supset V_2^n \supset \cdots \supset V_k^n.$$

The set V_i^n is a closed algebraic or analytic set of codimension i, for $i = 0, \ldots, k$ and $V_{i-1}^n \setminus V_i^n$ is a regular manifold of codimension $i - 1$, for all $i = 1, \ldots, k$. Each $X \in V_{i-1}^n \setminus V_i^n$ is V_{i-1}^n-stable. Namely, there exists an open neighborhood U of X in G^n such that, if $Y \in U \cap (V_{i-1}^n \setminus V_i^n)$, then Y is C^0 equivalent to X.

A germ X belongs to the stratum V_1^n if at least one eigenvalue of $DX(0)$ has real part equal zero. Hence $V_0^n \setminus V_1^n$ consists of hyperbolic singularities and stability in $V_0^n \setminus V_1^n$ follows from the Hartman–Grobman Theorem. Subsequent strata are characterized by extra degeneracies, either at the level of the 1-jet or at the level of higher order expansions.

In Takens (1974b), Takens provided the classification of singularities with codimension less or equal than two for arbitrary n:

$$G^n = V_0^n \supset V_1^n \supset V_2^n \supset V_3^n, \tag{5.1}$$

with respect to a weak notion of C^0 equivalence where one only requires that the phase portraits reduced to the stable and unstable sets are preserved. Given a vector field with $X(0) = 0$, the stable (resp. unstable) set of 0 consists of all points $x \in \mathbb{R}^n$ whose forward (resp. backward) orbit tends to 0 as t tends to ∞ (resp. $-\infty$).

Takens classification was extended to singularities up to codimension four in Dumortier (1977) for planar vector fields:

$$G^2 = V_0^2 \supset V_1^2 \supset V_2^2 \supset V_3^2 \supset V_4^2 \supset V_5^2,$$

and in Dumortier and Ibáñez (1996, 1998, 1999) for three-dimensional vector fields:

$$G^3 = V_0^3 \supset V_1^3 \supset V_2^3 \supset V_3^3 \supset V_4^3 \supset V_5^3.$$

The notion of C^0 equivalence is used in both extensions. As expected, the stratifications provided in Dumortier and Ibáñez (1996, 1998, 1999) and Dumortier (1977) up to codimension two are the same as in Takens (1974b). We remark that a semialgebraic classification is not possible in G^3 at the level of singularities of codimension four Dumortier and Ibáñez (1998).

One of the ultimate goals of Bifurcation Theory is to elucidate the different dynamics that can emerge close to singularities at the different levels of stratification. To do that, given a singularity of codimension k, that is, a singularity in $V_k^n \setminus V_{k+1}^n$, one has to study its generic k-parameter unfoldings. Genericity means that the family intersects transversely the stratum of codimension k.

Remark 5.1. *Although all classification results are stated for C^∞ vector fields, they are also valid for C^r vector fields if r is big enough.*

Remark 5.2. *Reduction to a center manifold, reduction to a normal form and blowing-up are three basic techniques not only for the classification of singularities, but also to study their unfoldings. They will be illustrated later on.*

5.2   Center manifolds and normal forms

In the next section we will present the two elementary bifurcations of codimension one: the saddle-node bifurcation and the Hopf bifurcation. For a greater generality we will refer to the reduction to the center manifold and we will also use the reduction of a singularity to a normal form. These are two essential techniques in Bifurcation Theory. Due to this reason, in this section we recall very briefly the main results about such techniques.

5.2.1   Center manifolds

When working with partially hyperbolic singularities, a reduction to a center manifold is required.

Consider a system

$$\begin{cases} \dot{x} &= Ax + F(x,y) \\ \dot{y} &= By + G(x,y), \end{cases} \tag{5.2}$$

where $x \in \mathbb{R}^p$ and $y \in \mathbb{R}^q$, A (resp. B) is a matrix whose eigenvalues have non-zero (resp. zero) real part, $F \in C^{k+1}(\mathbb{R}^{p+q}, \mathbb{R}^p)$, $G \in C^{k+1}(\mathbb{R}^{p+q}, \mathbb{R}^q)$, with $k \geqslant 1$, and both F and G are of order $o(\|(x,y)\|)$.

Theorem 5.3 (Carr (1982)). *There exists $\varepsilon > 0$ and a C^k function*

$$h : \{u \in \mathbb{R}^q : \|u\| < \varepsilon\} \to \mathbb{R}^p$$

with $h(0) = 0$ and $Dh(0) = 0$, such that the manifold

$$W^c = \{(h(u), u) : u \in \mathbb{R}^q, \|u\| < \varepsilon\}$$

is invariant under the flow of system (5.2).

W^c is said to be a center manifold of the equilibrium $(0,0) \in \mathbb{R}^{p+q}$. Center manifolds may not be unique, but their tangent space at the origin is uniquely determined by the eigenspace associated with eigenvalues with real part equal zero.

Theorem 5.4 (Kirchgraber and Palmer (1990)). *System (5.2) is C^0 equivalent (in a neighborhood of $(0,0) \in \mathbb{R}^{p+q}$) to*

$$\begin{cases} \dot{u} &= Au \\ \dot{v} &= Bv + G(h(v), v) \end{cases} \tag{5.3}$$

with $(u, v) \in \mathbb{R}^p \times \mathbb{R}^q$ and h is as given in Theorem 5.3.

Second equation in system (5.3) is said to be the reduction to center manifold of system (5.2). Note that both results can be applied to families of vector fields because, given a family X_λ defined in a neighborhood of $0 \in \mathbb{R}^n$ for parameter values λ in a neighborhood of $0 \in \mathbb{R}^k$ and with $X_0(0) = 0$, one can apply Theorem 5.3 and Theorem 5.4 to the system

$$\begin{cases} \dot{x} &= X_\lambda(x) \\ \dot{\lambda} &= 0. \end{cases}$$

Remark 5.5. *It must be remarked that even if a vector field is C^∞, associated center manifolds do not have to be C^∞ van Strien (1979), but there will exist a C^k center manifold for any given k. One-dimensional center manifolds are an exception, but regularity is not essential for our discussion and so, we do not enter in more details. An extended discussion can be found for instance in Broer, Dumortier, et al. (1991).*

5.2.2 Normal forms

Let $X = L + F$ be a vector field with L linear and F a C^r vector field such that $F(0) = 0$ and $DF(0) = 0$.

The Normal Form Theory seeks to obtain simplified expressions of a vector field in the neighborhood of an equilibrium point. With that goal, the following adjoint action is considered:

$$\mathrm{ad}\, L : \Psi \rightarrow [L, \Psi], \tag{5.4}$$

where Ψ is a vector field and $[L, \Psi]$ is also a vector field defined as

$$[L, \Psi](x) = D\Psi(x)L(x) - DL(x)\Psi(x). \tag{5.5}$$

Let $\mathscr{H}^m$ be the space of vector fields whose components are homogeneous polynomials of degree m. Denote by $\operatorname{ad}_m L$ the linear operator given by the restriction of $\operatorname{ad} L$ to $\mathscr{H}^m$ and write

$$\mathscr{H}^m = \mathscr{B}^m \oplus \mathscr{G}^m,$$

where $\mathscr{B}^m$ is the image of $\operatorname{ad}_m L$ and $\mathscr{G}^m$ is some complement.

Theorem 5.6. *Up to an analytic change of coordinates X can be written as*

$$X = L + g_2 + \cdots + g_r + R$$

where $g_i \in \mathscr{G}^i$ for each $i = 2, \ldots, r$ and $R(x) = o(\|x\|^r)$.

There are many references where a proof of Theorem 5.6 can be found. Simple arguments are given, for instance, in Guckenheimer and Holmes (2002), Kuznetsov (2004) or Broer, Dumortier, et al. (1991). See also the original proof can be seen in Takens (1974b).

Remark 5.7. *It follow from the Borel Theorem for representations of ∞-jets that theorem is valid for $r = \infty$ (see Broer, Dumortier, et al. (1991) for additional details). In Broer (1981) it is argued that there are structures, like volume-preserving properties or reversibility which can be preserved by reduction to normal form.*

The Normal Form Theorem can be extended to unfoldings of singularities. The simplest approach is to consider a reduction based in the decomposition of $\mathscr{H}^m$ by the image of $\operatorname{ad}_m L$ and some complement $\mathscr{G}^m$, but assuming that coefficients in the polynomial expressions depend on parameters. This setting is enough for our purposes. More elaborated results are provided in Broer, Dumortier, et al. (1991).

5.3   Some elementary singularities

Given a n-dimensional singularity X, that is, a vector field in $\mathbb{R}^n$ such that $X(0) = 0$, and a family of vector fields X_λ, on $\mathbb{R}^n$, with $\lambda \in \mathbb{R}^k$, such that $X_0 = X$, we say that X_λ is an unfolding of the singularity.

5.3.1   Saddle-node bifurcation

Consider a C^∞ vector field X defined in a neighborhood of $0 \in \mathbb{R}^n$ such that $X(0) = 0$ and $DX(0)$ has eigenvalues

$$\{0, \mu_1, \ldots, \mu_p, \nu_1, \ldots, \nu_q\},$$

with $p + q = n - 1$, $\operatorname{Re} \mu_i < 0$ for all $i = 1, \ldots, p$ and $\operatorname{Re} \nu_i > 0$ for all $i = 1, \ldots, q$. Note that X belongs to the stratum V_1^n in the Takens classification (5.1).

As already discussed in §5.2, in this case we know that there exists a one-dimensional invariant manifold W^c, the center manifold, which is tangent at 0 to the eigenspace associated with the zero eigenvalue. Moreover, X is C^0 conjugate to the vector field

$$\widetilde{X}(x)\frac{\partial}{\partial x} - \sum_{i=1}^{p} y_i \frac{\partial}{\partial y_i} + \sum_{i=1}^{q} z_i \frac{\partial}{\partial z_i}$$

with $\widetilde{X}(0) = 0$ and $\widetilde{X}'(0) = 0$. The one-dimensional vector field

$$\widetilde{X}(x)\frac{\partial}{\partial x} \tag{5.6}$$

is the reduction of X to the center manifold and it is given by the restriction of X to the invariant manifold W^c. As already mentioned in §5.2, smoothness of the reduction is always a delicate point, but when W^c is one-dimensional, it is known that $\widetilde{X}$ is C^∞ (see comments in Broer, Dumortier, et al. (1991) and Kuznetsov and Meijer (2019) and also a nice proof in Dumortier, Llibre, and Artés (2006)). Therefore, to understand the local behaviour around this n-dimensional singularity, one only needs to study a one-dimensional vector field.

In the sequel we assume that $\widetilde{X}''(0) \neq 0$. This condition characterizes a singularity of codimension one. It is straightforward that (5.6) is locally C^0 equivalent to the vector field

$$u^2 \frac{\partial}{\partial u} \tag{5.7}$$

and therefore, there is a unique topological type. Figure 5.1 shows the (trivial) behaviour exhibited by the flow of (5.7).

Figure 5.1: *Phase-portrait of the vector field in (5.7).*

In the sequel we restrict to the one-dimensional case. Let $\widetilde{X}_\lambda$ be a C^∞ k-parameter family such that $\widetilde{X}_0 = \widetilde{X}$, that is, an unfolding of the singularity (5.6). It can be written as:

$$\widetilde{X}_\lambda : \left(a_0(\lambda) + a_1(\lambda)u + a_2(\lambda)u^2 + r(\lambda, u) \right) \frac{\partial}{\partial u} \tag{5.8}$$

where a_0, a_1 and a_2 are C^∞ functions and $r(\lambda, u) = O(|u|^3)$. Taking into account the conditions defining the singularity, it follows that $a_0(0) = a_1(0) = 0$ and $a_2(0) \neq 0$.

Applying the Malgrange Preparation Theorem (see Chow and Hale (1982)) we know that there exists a C^∞ function $\varphi(\lambda, u)$ with $\varphi(0,0) > 0$ such that

$$a_0(\lambda) + a_1(\lambda)u + a_2(\lambda)u^2 + r(\lambda, u) = \varphi(\lambda, u)\left(\hat{a}_0(\lambda) + \hat{a}_1(\lambda)u \pm u^2\right).$$
$$(5.9)$$

Clearly, identity provides a C^∞-equivalence between $\widetilde{X}_\lambda$ and

$$\widehat{X}_\lambda \; : \; \left(\hat{a}_0(\lambda) + \hat{a}_1(\lambda)u \pm u^2\right)\frac{\partial}{\partial u}. \qquad (5.10)$$

Moreover, introducing $\bar{u} = -u$ we can reduce the study to the case with $+$ sign. Now we consider the change of coordinates $\bar{u} = u + \hat{a}_1(\lambda)/2$ to obtain the C^∞-equivalent family

$$Y_\lambda \; : \; \left(\tilde{a}_0(\lambda) + \bar{u}^2\right)\frac{\partial}{\partial \bar{u}}. \qquad (5.11)$$

Finally, assuming the generic hypothesis $\tilde{a}_0'(0) \neq 0$, we can introduce a new parameter $\mu = \tilde{a}_0(\lambda)$ to get the μ-dependent family:

$$\left(\mu + \bar{u}^2\right)\frac{\partial}{\partial \bar{u}}. \qquad (5.12)$$

Figure 5.2 shows the bifurcation diagram associated with the above family.

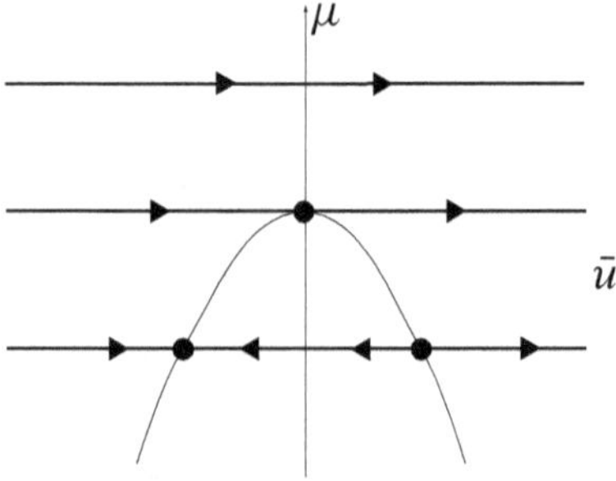

Figure 5.2: *Saddle-node bifurcation.*

5.3.2 Hopf bifurcation

Let X be a C^∞ vector field defined in a neighborhood of $0 \in \mathbb{R}^n$ such that $X(0) = 0$ and $DX(0)$ has eigenvalues

$$\{\omega i, -\omega i, \mu_1, \ldots, \mu_p, \nu_1, \ldots, \nu_q\},$$

with $p + q = n - 2$, $\omega \neq 0$, $\operatorname{Re} \mu_i < 0$ for all $i = 1, \ldots, p$ and $\operatorname{Re} \nu_i > 0$ for all $i = 1, \ldots, q$. It easily follows that inside the space of $n \times n$ matrices, the subset given by those with a pair of pure imaginary eigenvalues has codimension one. Hence, bearing in mind the Takens classification (5.1), the germ of X belongs to the stratum V_1^n.

In this case, we know that there exists a two-dimensional invariant manifold W^c, the center manifold, which is tangent at 0 to the eigenspace associated with the pair of imaginary eigenvalues. Moreover, X is C^0 conjugate to the vector field

$$\omega \left(u \frac{\partial}{\partial v} - v \frac{\partial}{\partial u} \right) + F(u, v) \frac{\partial}{\partial u} + G(u, v) \frac{\partial}{\partial v} - \sum_{i=1}^{p} y_i \frac{\partial}{\partial y_i} + \sum_{i=1}^{q} z_i \frac{\partial}{\partial z_i},$$

with $F(0,0) = G(0,0) = 0$ and $DF(0,0) = DG(0,0) = 0$. The two-dimensional vector field

$$R = \omega \left(u \frac{\partial}{\partial v} - v \frac{\partial}{\partial u} \right) + F(u, v) \frac{\partial}{\partial u} + G(u, v) \frac{\partial}{\partial v} \tag{5.13}$$

is the reduction of X to the center manifold and it is given by the restriction of X to the invariant manifold W^c. Vector field R can be assumed to be C^r with r arbitrarily large. To understand the local behaviour around this n-dimensional singularity, we need to study the planar vector field R.

One can compute a normal form to get an equivalent but simplified expression of R. With this regard, it is more convenient to introduce complex coordinates $z = u + iv$ and $\bar{z} = u - iv$ to transform (5.13) into:

$$\widetilde{R} = \omega i \left(z \frac{\partial}{\partial z} - \bar{z} \frac{\partial}{\partial \bar{z}} \right) + \widetilde{F}(z, \bar{z}) \frac{\partial}{\partial z} + \widetilde{G}(z, \bar{z}) \frac{\partial}{\partial \bar{z}}. \tag{5.14}$$

with $\widetilde{F}(0,0) = \widetilde{G}(0,0) = 0$ and $D\widetilde{F}(0,0) = D\widetilde{G}(0,0) = 0$. The adjoint action of the linear part

$$L = \left(z \frac{\partial}{\partial z} - \bar{z} \frac{\partial}{\partial \bar{z}} \right) \tag{5.15}$$

on elementary monomials is simple:

$$\left[L, z^i \bar{z}^j \frac{\partial}{\partial z} \right] = (i - j - 1) z^i \bar{z}^j \frac{\partial}{\partial z}$$

$$\left[L, z^i \bar{z}^j \frac{\partial}{\partial \bar{z}} \right] = (i - j + 1) z^i \bar{z}^j \frac{\partial}{\partial \bar{z}}.$$

It easily follows that for each degree $m \geq 2$ fixed, all monomials with $z^i \bar{z}^{m-i} \frac{\partial}{\partial z}$, with $2i - m - 1 \neq 0$, and $z^i \bar{z}^{m-i} \frac{\partial}{\partial \bar{z}}$, with $2i - m + 1 \neq 0$, are removable by an analytical change of coordinates. Hence, one can assume that

$$\widetilde{R} = \omega \mathrm{i} \left(z \frac{\partial}{\partial z} - \bar{z} \frac{\partial}{\partial \bar{z}} \right)$$
$$+ z \left(\sum_{s=1}^{k} a_s (z\bar{z})^s \right) \frac{\partial}{\partial z} + \bar{z} \left(\sum_{s=1}^{k} \bar{a}_s (z\bar{z})^s \right) \frac{\partial}{\partial \bar{z}} + o(\|(z,\bar{z})\|^{2k+1}) \tag{5.16}$$

with k arbitrarily large and $a_s = \alpha_s + i\beta_s$.

Recovering cartesian coordinates we obtain:

$$\bar{R} = \left(\omega + \sum_{s=1}^{k} \beta_s (u^2 + v^2)^s \right) \left(u \frac{\partial}{\partial v} - v \frac{\partial}{\partial u} \right)$$
$$+ \left(\sum_{s=1}^{k} \alpha_s (u^2 + v^2)^s \right) \left(u \frac{\partial}{\partial u} + v \frac{\partial}{\partial v} \right) + o \left(\|(u,v)\|^{2k+1} \right). \tag{5.17}$$

The generic condition $\alpha_1 \neq 0$ characterizes a singularity of codimension one with two topological types which depend on the sign of α_1 (see Figure 5.3).

Consider now a C^r-unfolding R_λ, with $\lambda \in \mathbb{R}^k$ and $k \geq 1$ such that $R_0 = R$, with R as given in (5.13). Because $\det DR(0) = \omega^2 \neq 0$, the singularity persists for small enough values of λ and we can assume that $R_\lambda(0) = 0$ for all λ. On the other hand, again for small λ, eigenvalues of $DR_\lambda(0)$ are of the form $\rho(\lambda) \pm i\tau(\lambda)$, with $\tau(0) = \omega \neq 0$. Without loss of generality, we can also assume that

$$DR_\lambda(0) = \begin{pmatrix} \rho(\lambda) & -\tau(\lambda) \\ \tau(\lambda) & \rho(\lambda) \end{pmatrix}.$$

Finally, we can redo a reduction to normal form, but in this case for the family. In any case, the adjoint action that we have to use is the same one that we have already considered for the reduction of the singularity itself, that is, the action of (5.15). It easily follows that we can write R_λ as:

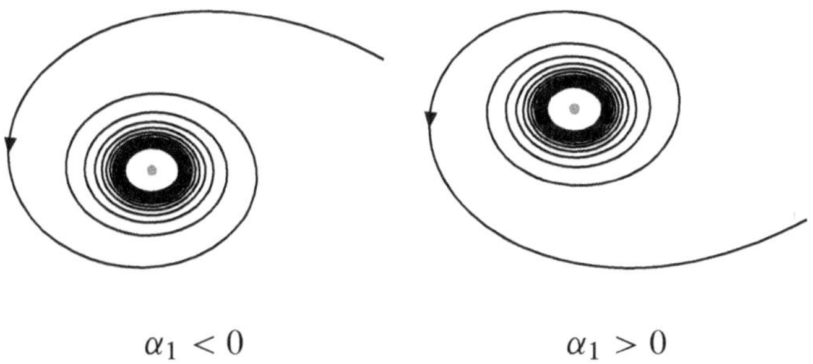

Figure 5.3: *Local topological types of (5.17) when $\alpha_1 \neq 0$.*

$$\overline{R}_\lambda = \left(\tau(\lambda) + \sum_{s=1}^{k} \beta_s(\lambda)(u^2 + v^2)^s\right)\left(u\frac{\partial}{\partial v} - v\frac{\partial}{\partial u}\right)$$

$$+ \left(\rho(\lambda) + \sum_{s=1}^{k} \alpha_s(\lambda)(u^2 + v^2)^s\right)\left(u\frac{\partial}{\partial u} + v\frac{\partial}{\partial v}\right) \tag{5.18}$$

$$+ o\left(\|(u,v)\|^{2k+1}\right).$$

Now we impose $\nabla\rho(0) \neq 0$. Without loss of generality one can assume that $\frac{\partial \rho}{\partial \lambda_1}(0) \neq 0$. Hence we can consider $(\rho, \lambda_2, \ldots, \lambda_n)$ as new parameters. Introducing polar coordinates in (5.17) one can easily determine the phase portrait of (5.16), but now we only refer to Kuznetsov (2004, Chapter 3) or Broer, Dumortier, et al. (1991, Chapter 7) to state that, for $(\lambda_2, \ldots, \lambda_k)$ small and fixed, the above family is locally C^0 equivalent to

$$u\frac{\partial}{\partial v} - v\frac{\partial}{\partial u} + \left(\rho + \kappa(u^2 + v^2)\right)\left(u\frac{\partial}{\partial u} + v\frac{\partial}{\partial v}\right). \tag{5.19}$$

with $\kappa = \pm 1$. Bifurcation diagrams for (5.19) are depicted in Figure 5.4. If $\kappa = -1$ (resp. $\kappa = +1$), there is an attracting (repelling) periodic orbit when $\rho > 0$ (resp. $\rho < 0$). This periodic orbit collapses with the singularity when $\rho = 0$ and disappears, transferring its character, either attracting or repelling, to the equilibrium point. The case $\kappa = -1$ (resp. $\kappa = +1$) is said supercritical (resp. subcritical) Hopf bifurcation.

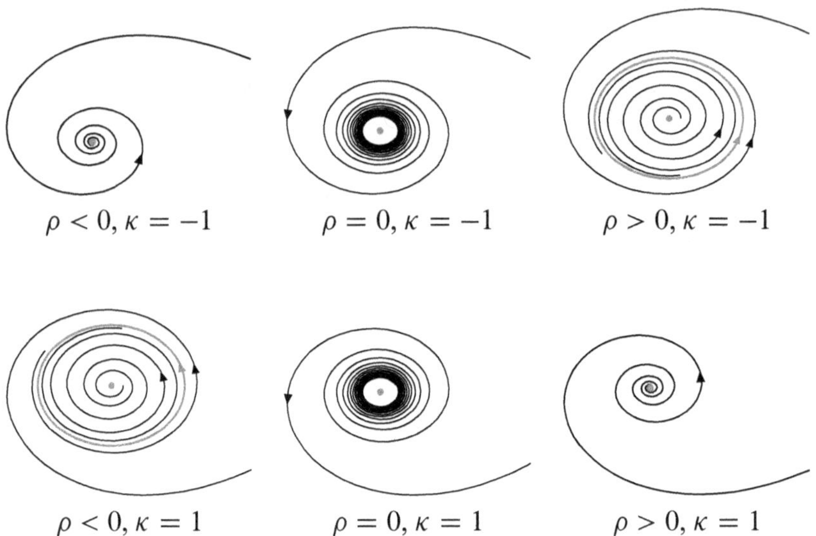

Figure 5.4: *Bifurcation diagrams for a Hopf bifurcation.*

5.4   Nilpotent singularities

Consider a C^∞ vector field X defined in a neighborhood of $0 \in \mathbb{R}^n$ such that $X(0) = 0$ and $DX(0)$ is linearly conjugate to $\sum_{k=1}^{n-1} x_{k+1} \frac{\partial}{\partial x_k}$. As argued in Drubi, Ibáñez, and Rodríguez (2007), introducing appropriate C^∞ coordinates, X can be written as

$$\sum_{k=1}^{n-1} x_{k+1} \frac{\partial}{\partial x_k} + f(x_1, \ldots, x_n) \frac{\partial}{\partial x_n}, \tag{5.20}$$

where $f(x_1, \ldots, x_n) = O(\|(x_1, \ldots, x_n)\|^2)$. If $\frac{\partial f}{\partial x_1^2}(0) \neq 0$ we refer to X as the *n-dimensional nilpotent singularity of codimension n*. The 2-dimensional nilpotent singularity of codimension two, also named *Bogdanov–Takens singularity*, was already classified in Takens (1974b), whereas the 3-dimensional nilpotent singularity of codimension three was classified in Dumortier and Ibáñez (1996). In both cases there is a unique topological type. Classification of n-dimensional nilpotent singularities of codimension n with $n \geq 4$ remains as an open problem.

According to Drubi, Ibáñez, and Rodríguez (2007), any generic n-parameter unfolding of a n-dimensional nilpotent singularity of codimension n can be written as

$$\sum_{k=1}^{n-1} x_{k+1}\frac{\partial}{\partial x_k} + \left(\mu_1 + \sum_{k=2}^{n} \mu_k x_k + x_1^2 + h(x,\mu)\right)\frac{\partial}{\partial x_n}, \qquad (5.21)$$

where $\mu = (\mu_1,\ldots,\mu_n) \in \mathbb{R}^n$, $x = (x_1,\ldots,x_n) \in \mathbb{R}^n$ and

$$h(0,\mu) = 0, \quad \frac{\partial h}{\partial x_i}(0,\mu) = 0 \quad \text{for} \quad i = 1,\ldots,n, \quad \frac{\partial^2 h}{\partial x_1^2}(0,\mu) = 0,$$

$$h(x,\mu) = O(\|(x,\mu)\|^2), \quad h(x,\mu) = O(\|(x_2,\ldots,x_n)\|).$$

Of course, the required generic conditions involve derivatives of the family with respect to parameters.

Techniques of reduction to normal allows to get further reductions in expression (5.21), but we do not need them for our purposes. Normal forms in the 2-dimensional case were already derived in Takens (1974a,b) (see also Broer, Dumortier, et al. (1991)). Normal forms in the 3-dimensional case are considered in detail in Dumortier and Ibáñez (1996).

Extending the blowing-up techniques used in Dumortier and Ibáñez (1996) for dimension three, we can rescale variables and parameters by means of

$$\begin{aligned}
\mu_1 &= \varepsilon^{2n} v_1 \\
\mu_k &= \varepsilon^{n-k+1} v_k \qquad \text{for} \quad k = 2,\ldots,n, \qquad (5.22)\\
x_k &= \varepsilon^{n+k-1} y_k \qquad \text{for} \quad k = 1,\ldots,n,
\end{aligned}$$

with $\varepsilon > 0$ and $v_1^2 + \cdots + v_n^2 = 1$ to write, after dividing by ε, (5.21) as

$$\sum_{k=1}^{n-1} y_{k+1}\frac{\partial}{\partial y_k} + \left(v_1 + \sum_{k=2}^{n} v_k y_k + y_1^2 + \varepsilon\kappa y_1 y_2 + O(\varepsilon^2)\right)\frac{\partial}{\partial y_n} \quad (5.23)$$

where

$$\kappa = \frac{\partial^2 h}{\partial x_1 \partial x_2}(0,0)$$

and $(y_1,\ldots,y_n) \in K$, where $K \subset \mathbb{R}^n$ is an arbitrarily big compact. As an extra non-degeneracy condition we assume that $\kappa \neq 0$.

When one uses rescaling techniques, the first step to study the dynamics arising in the unfolding of a singularity is to understand the dynamics exhibited by the limit family obtained when $\varepsilon = 0$. Namely, we should be interested in the family:

$$\sum_{k=1}^{n-1} y_{k+1}\frac{\partial}{\partial y_k} + \left(v_1 + \sum_{k=2}^{n} v_k y_k + y_1^2\right)\frac{\partial}{\partial y_n}. \qquad (5.24)$$

In subsequent sections we will extend on details about the unfoldings in the cases of phase space with dimensions 2, 3 and 4. Here we establish some generalities to which we will refer later.

When $v_1 > 0$ there are no singularities. In fact, dynamical behaviour for (5.24) is rather simple under such restriction because, as one can easily check, the function

$$L(y_1, \ldots, y_n) = y_n - \sum_{k=1}^{n-1} v_{k+1} y_k.$$

is strictly increasing along the orbits and, as a consequence, the maximal compact invariant set is empty. Even more, when $v_1 = 0$,

$$\frac{dL}{dt} = y_1^2 \geq 0.$$

and hence limit sets must be contained in the plane $y_1 = 0$. Since the maximal invariant set contained in $\{y_1 = 0\}$ consists only of the equilibrium point $(0,0)$ then any forward bounded orbit must have ω-limit $\{(0,0)\}$. Similarly, any backward bounded orbit has α-limit $\{(0,0)\}$. We conclude that one only needs to study the case $v_1 < 0$.

Family (5.24) also exhibits some symmetries. Namely, up to a change of sign, it is invariant with respect to the transformation:

$$(v, y) \quad \rightarrow \quad \big(v_1, (-1)^{n-1} v_2, (-1)^{n-2} v_3, \ldots, v_{n-1}, -v_n,$$
$$(-1)^n y_1, (-1)^{n-1} y_2, \ldots, y_{n-1}, -y_n\big).$$

It follows that we can consider a reduced set of parameters with $v_1 < 0$ and $v_n \leq 0$. Moreover, for parameter values on the set

$$\mathscr{T} = \{(v_1, \ldots, v_n) \in \mathbb{S}^{n-1} : v_{n-2i} = 0 \text{ with } i = 0, \ldots, \lfloor (n-2)/2 \rfloor\}$$

system (5.24) is time-reversible with respect to the involution

$$R : (y_1, \ldots, y_n) \rightarrow \big((-1)^n y_1, (-1)^{n-1} y_2, \ldots, y_{n-1}, -y_n\big).$$

The manifold $\mathscr{T}$ has dimension $\lfloor (n-1)/2 \rfloor$ and we refer to it as the reversibility set of the limit family.

To conclude this section about generalities regarding the unfolding of n-dimensional nilpotent singularities of codimension n, we observe that the divergence of the limit family equals v_n. Hence, if $v_n = 0$, family (5.24) is volume-preserving. In fact, assuming that n is even, one can introduce appropriate coordinates to write the family as a Hamiltonian system. Namely, the following result is proved in Barrientos, Ibáñez, and Rodríguez (2011):

Theorem 5.8. *Introducing the new variables*

$$q = S \cdot (y_1, y_3, \ldots, y_{n-1})^t, \qquad p = (y_2, y_4, \ldots, y_n)^t,$$

with

$$\begin{pmatrix} -v_3 & -v_5 & \cdots & -v_{n-1} & 1 \\ -v_5 & & \cdot^{\cdot^\cdot} & \cdot^{\cdot^\cdot} & 0 \\ \vdots & \cdot^{\cdot^\cdot} & \cdot^{\cdot^\cdot} & \cdot^{\cdot^\cdot} & \vdots \\ -v_{n-1} & \cdot^{\cdot^\cdot} & \cdot^{\cdot^\cdot} & & \vdots \\ 1 & 0 & \cdots & \cdots & 0 \end{pmatrix}$$

the family

$$\sum_{k=1}^{n-1} y_{k+1} \frac{\partial}{\partial y_k} + \left(v_1 + \sum_{k=1}^{m-1} v_{2k+1} y_{2k+1} + y_1^2 \right) \frac{\partial}{\partial y_n}.$$

transforms into

$$\frac{\partial H}{\partial p} \frac{\partial}{\partial q} - \frac{\partial H}{\partial q} \frac{\partial}{\partial p}$$

where

$$H(p, q) = \frac{1}{2} \langle Sp, p \rangle + V(q).$$

The potential V is defined as

$$V(q) = -\frac{1}{3} q_m^3 - \frac{1}{2} \sum_{k=1}^{m-1} v_{2k+1} b_{k+1} q_m^2 - \frac{1}{2} \sum_{j=1}^{\lfloor m/2 \rfloor} b_{m-2j+1} q_{m-j}^2$$

$$- \sum_{k=1}^{m-1} \sum_{i=m-k}^{m-1} v_{2k+1} b_{i-m+k+1} q_i q_m - \sum_{j=1}^{\lfloor m/2 \rfloor} \sum_{i=j}^{m-j-1} b_i q_i q_{m-j} - v_1 q_m$$

where, given $b_1 = 1$,

$$b_i = \sum_{l=1}^{i-1} v_{2(m-i+l)+1} b_l \qquad for\ i = 2, \ldots, m.$$

Regarding Hamiltonian vector fields, the following result will be useful in the next sections.

Proposition 5.9. *Given a enough smooth function $H : \mathbb{R}^{2n} \to \mathbb{R}$, consider the Hamiltonian vector field $X_H(u) = J\nabla H(u)$ where J is the Poisson matrix, that is,*

$$J = \begin{pmatrix} 0 & -I \\ I & 0 \end{pmatrix}$$

where I is the $n \times n$ identity matrix. Let $p(t)$ be a solution of the Hamiltonian equation $\dot{u} = X_H(u)$. Then $\psi(t) = \nabla H(p(t))$ is a solution of the adjoint variational equation $\dot{w} = -DX_H(p(t))w$.

Proof. Note that

$$\dot{\psi}(t) = D^2 H(p(t))\dot{p}(t) = D^2 H(p(t))J\nabla H(p(t))$$
$$= -D^2 H(p(t))J^*\psi(t) = -DX_H(p(t))^*\psi(t). \qquad \square$$

5.5   Nilpotent of codimension two in $\mathbb{R}^2$

Consider C^∞ vector fields X defined in a neighborhood of $0 \in \mathbb{R}^2$ such that $X(0) = 0$ and $DX(0)$ is linearly conjugate to

$$x_2 \frac{\partial}{\partial x_1}. \tag{5.25}$$

As explained in §5.4, X can be written as

$$x_2 \frac{\partial}{\partial x_1} + f(x) \frac{\partial}{\partial x_2},$$

where $x = (x_1, x_2)$ and $f(x) = O(\|x\|^2)$. The condition $\frac{\partial^2 f}{\partial x_1^2}(0) \neq 0$ characterizes the two-dimensional nilpotent singularity of codimension two. As follows from Takens (1974b) (see also Broer, Dumortier, et al. (1991)), two-dimensional nilpotent singularities exhibit a unique topological type (see Figure 5.5).

Any generic two-parameter unfolding of this singularity can be written as

$$x_2 \frac{\partial}{\partial x_1} + \left(\mu_1 + \mu_2 x_2 + x_1^2 + h(x, \mu)\right) \frac{\partial}{\partial x_2}, \tag{5.26}$$

where $\mu = (\mu_1, \mu_2)$, $h(0, \mu) = 0$, $\frac{\partial h}{\partial x_i}(0, \mu) = 0$, for $i = 1, 2$, $\frac{\partial^2 h}{\partial x_1^2}(0, \mu) = 0$, $h(x, \mu) = O(\|(x, \mu)\|^2)$ and also $h(x, \mu) = O(\|(x_2)\|)$. The local bifurcation diagram for family (5.26) was determined independently by Bogdanov (1975) and Takens (1974a) and unfoldings of the two-dimensional nilpotent singularity of codimension two are called Bogdanov–Takens bifurcations. Given an

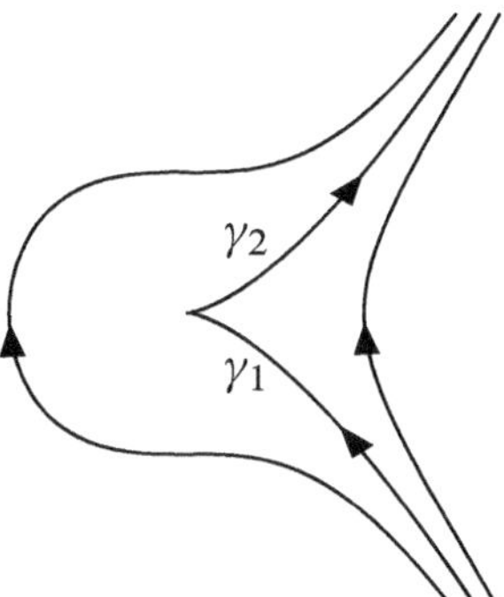

Figure 5.5: *Topological type of the Bogdanov–Takens singularity. Orbits γ_1 and γ_2 are such that $\omega(\gamma_1) = \{(0,0)\}$ and $\alpha(\gamma_2) = \{(0,0)\}$. Excepting these two orbits, the phase-portrait shows a flow-box around the singularity.*

arbitrary unfolding (5.26), there exists a local homeomorphism in the phase-space, depending continuously with respect to the parameters, and also a local homeomorphism in the parameter space such that family (5.26) changes into (see Broer, Dumortier, et al. (1991)):

$$x_2 \frac{\partial}{\partial x_1} + \left(\mu_1 + \mu_2 x_2 + x_1^2 + x_1 x_2\right) \frac{\partial}{\partial x_2}. \tag{5.27}$$

Figure 5.6 provides and sketch of the bifurcation diagram for family (5.27).

As already explained, the Bogdanov–Takens bifurcation is well-known from the literature. Nevertheless it is a quite illustrative example of part of the ideas used in higher-dimensions to prove the existence of homoclinic orbits or heteroclinic cycles. Due to that reason, we include here the proof of the existence of homoclinic bifurcation curve in any generic unfolding of the two-dimensional nilpotent singularity of codimension two.

Theorem 5.10. *There exists a curve* Hom *(see Figure 5.6) in* $\mathbb{R}^2$ *with an end at the origin such that for each* $(\mu_1, \mu_2) \in$ Hom, *the system (5.26) exhibits a homoclinic orbit.*

Blowing-up coordinates and parameters in (5.26) as done in (5.22)

$$\begin{aligned} \mu_1 &= \varepsilon^4 v_1, & \mu_2 &= \varepsilon v_2, \\ x_1 &= \varepsilon^2 y_1, & x_2 &= \varepsilon^3 y_2, \end{aligned} \tag{5.28}$$

we get, after rescaling time by a factor ε, the expression (5.23) for $n = 2$,

$$y_2 \frac{\partial}{\partial y_1} + \left(v_1 + v_2 y_2 + y_1^2 + \varepsilon \kappa y_1 y_2 + O(\varepsilon^2)\right) \frac{\partial}{\partial y_4}, \tag{5.29}$$

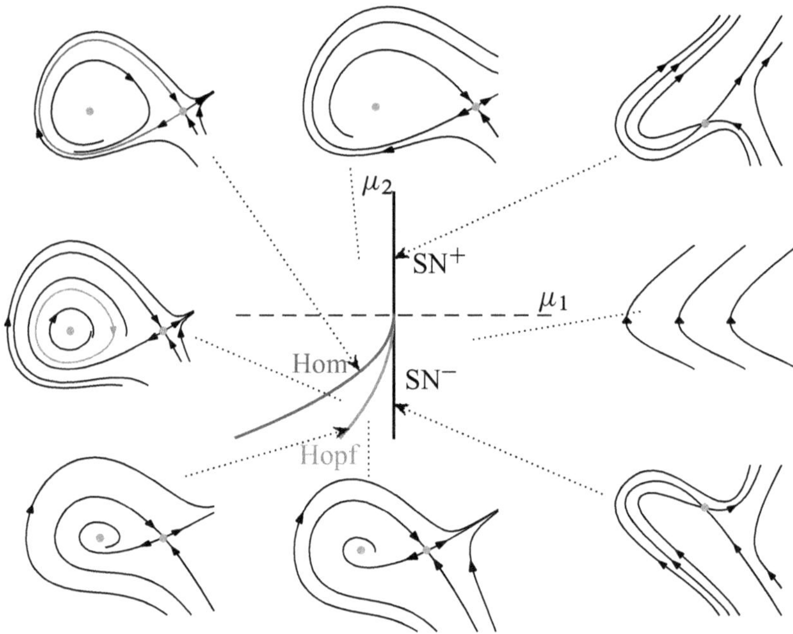

Figure 5.6: *Bogdanov–Takens bifurcation. Three codimension one bifurcations are unfolded. Vertical axis is a line of saddle-bifurcation. For parameters along the positive (resp. negative) vertical axis, labelled SN^+ (resp. SN^-), the equilibrium point has a one-dimensional center manifold and a one-dimensional unstable (resp. stable) invariant manifold. Red and blue curves corresponds to Hopf bifurcation and homoclinic bifurcation, respectively. For parameters in between these curves, the system exhibits a unique attracting limit cycle (plotted in red in the corresponding phase-portrait). Homoclinic loop is plotted in blue colour in the phase-portrait corresponding to the homoclinic bifurcation curve. Equilibrium points are plotted in magenta.*

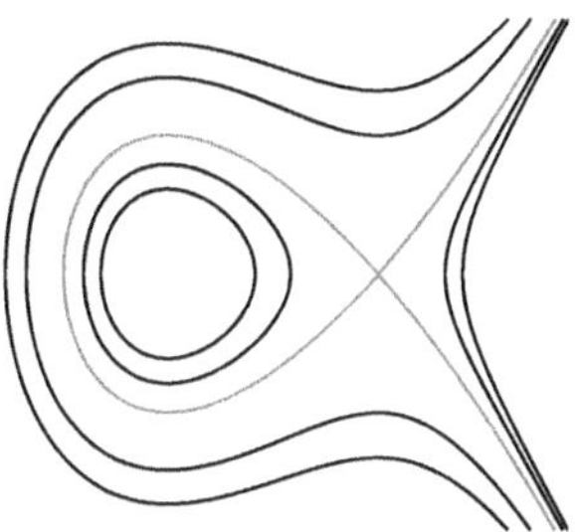

Figure 5.7: *Level curves of H.*

where we assume that

$$\kappa = \frac{\partial^2 h}{\partial x_1 \partial x_2}(0,0) \neq 0.$$

As usual, we consider $y = (y_1, y_2)$ varying in an arbitrarily big compact in $\mathbb{R}^2$ and $v \in \mathbb{S}^1$. The first goal is to understand the dynamics exhibited by the limit family

$$y_2 \frac{\partial}{\partial y_1} + \left(v_1 + v_2 y_2 + y_1^2\right) \frac{\partial}{\partial y_2}, \tag{5.30}$$

obtained by taking $\varepsilon = 0$ in (5.29). Note that, as already discussed in §5.4, one only needs to consider $v_1 < 0$ and $v_2 \leqslant 0$.

When $v_2 = 0$ (5.30) is time-reversible with respect to the involution

$$R : (y_1, y_2) \to (y_1, -y_2).$$

Under the restriction $v_1 < 0$, it is more convenient to consider a directional rescaling of the parameters. We use the same formulas as in the spherical rescaling defined in (5.28), but we take $v_1 = -1$ and assume that $v_2 \in \mathbb{R}$ to obtain

$$y_2 \frac{\partial}{\partial y_1} + \left(-1 + v_2 y_2 + y_1^2 + \varepsilon \kappa y_1 y_2 + O(\varepsilon^2)\right) \frac{\partial}{\partial y_2}. \tag{5.31}$$

When $v_2 = 0$ and $\varepsilon = 0$ the system has the first integral:

$$H(y_1, y_2) = \frac{y_2^2}{2} - \frac{y_1^3}{3} + y_1.$$

Level curves of the Hamiltonian H are given in Figure 5.7.

There are two equilibrium points $P_\pm = (\pm 1, 0)$. The equilibrium P_- is a center whereas P_+ is a saddle. Note that there is a homoclinic orbit to P_+ (see

Figure 5.7) and we are interested in the existence of homoclinic orbits when $\varepsilon > 0$.

We will consider family (5.31) as an unfolding of the homoclinic orbit exhibited by the system when $v_2 = 0$ and $\varepsilon = 0$. Moreover, we consider the family (5.31) written as

$$\dot{x} = f(x) + g(\lambda, x) \tag{5.32}$$

where

$$\lambda = (\lambda_1, \lambda_2) = (v_2 + \kappa\varepsilon, \varepsilon)$$

varies in a neighborhood of $(0, 0) \in \mathbb{R}^2$,

$$x = (x_1, x_2) = (y_1 - 1, y_2)$$
$$f(x_1, x_2) = (x_2, 2x_1 + x_1^2)$$
$$g(x_1, x_2) = (0, \lambda_1 x_2 + \lambda_2 \kappa x_1 x_2 + O(\lambda_2^2)).$$

Note that there is a translation of coordinates to move the saddle from $(1, 0)$ to $(0, 0)$. Bifurcations exhibited when $\lambda_2 > 0$ correspond to bifurcations observed in any generic unfolding of the two-dimensional nilpotent singularity of codimension two.

We will prove that for parameter values along a curve through the origin, there exist homoclinic orbits, that is, the unfolding of the homoclinic connection exhibited for the system when $\lambda = 0$ inside family (5.31) is generic.

Note that the unperturbed system $\dot{x} = f(x)$ satisfies the following properties:

(H1) There is a first integral

$$\widehat{H}(x_1, x_2) = \frac{x_2^2}{2} - \frac{x_1^3}{3} - x_1^2.$$

(H2) The system is time reversible with respect to the involution

$$R : (x_1, x_2) \to (x_1, -x_2).$$

(H3) There exists a homoclinic orbit

$$\gamma = \{p(t) = (p_1(t), p_2(t)) : t \in \mathbb{R}\},$$

such that p_1 is and even function and p_2 is an odd function. Moreover $p_1(t) < 0$ and $p_2(t) > 0$ for all $t \in (0, \infty)$.

(H4) According to Definition A.22, the homoclinic orbit γ is non-degenerate and, as follows from Proposition A.15 and Proposition A.21, the dimension of the space of bounded solutions for the variational equation

$$\begin{pmatrix} \dot{z}_1 \\ \dot{z}_2 \end{pmatrix} = Df(p(t)) \begin{pmatrix} z_1 \\ z_2 \end{pmatrix} = \begin{pmatrix} 0 & 1 \\ 2 + 2p_1(t) & 0 \end{pmatrix} \begin{pmatrix} z_1 \\ z_2 \end{pmatrix}$$

and also for the adjoint variational equation

$$\begin{pmatrix} \dot{w}_1 \\ \dot{w}_2 \end{pmatrix} = -Df(p(t))^* \begin{pmatrix} w_1 \\ w_2 \end{pmatrix} = \begin{pmatrix} 0 & -2 - 2p_1(t) \\ -1 & 0 \end{pmatrix} \begin{pmatrix} w_1 \\ w_2 \end{pmatrix}$$

equals one. The function $f(p(t))$ is a bounded solution of the variational equation and, according to Proposition 5.9,

$$\psi(t) = \nabla \widehat{H}(p(t)) = (-2p_1(t) - p_1(t)^2, p_2(t)))$$

is a bounded solution of the adjoint variational equation.

To study the persistence of the homoclinic orbit γ we consider the splitting function $\xi^\infty : \Lambda \to \mathbb{R}$ as defined in Lemma B.1. For each λ, $\xi^\infty(\lambda)$ provides the displacement function between the one-dimensional invariant manifolds on $f(p(0))^\perp$, Hence, for each λ such that $\xi^\infty(\lambda) = 0$, there is a homoclinic orbit at the origin.

Let

$$D_\lambda \xi^\infty(0) = \nabla \xi^\infty(0) = \begin{pmatrix} \xi^\infty_{\lambda_1}(0) & \xi^\infty_{\lambda_2}(0) \end{pmatrix}.$$

The bifurcation equation $\xi^\infty(\lambda) = 0$ can be studied by means of the Implicit Function Theorem. Namely, if $\operatorname{rank} D_\lambda \xi^\infty(0) = 1$, then the system has homoclinic orbits for parameter values on a curve $\mathscr{H}$ containing $\lambda = 0$. According to Theorem B.2

$$\xi^\infty_{\lambda_i}(0) = \int_{-\infty}^{\infty} \left\langle \psi(t), \frac{\partial g}{\partial \lambda_i}(0, p(t)) \right\rangle dt$$

and it easily follows that

$$\xi^\infty_{\lambda_1}(0) = \int_{-\infty}^{\infty} p_2(t)^2 \, dt,$$

$$\xi^\infty_{\lambda_2}(0) = \kappa \int_{-\infty}^{\infty} p_1(t) p_2(t)^2 \, dt.$$

It is straightforward that $\xi^\infty_{\lambda_1}(0) \neq 0$. On the other hand, since p_1 is even and $p_1(t) < 0$ for all $t \in [0, \infty)$, then $\xi^\infty_{\lambda_2}(0) \neq 0$.

As $\nabla \xi^\infty(0) \neq (0, 0)$, there exists a curve $\mathscr{H}$ corresponding to parameter values for which the systems exhibits a homoclinic orbit. Moreover, because $\xi^\infty_{\lambda_1}(0) \neq 0$, $\mathscr{H}$ intersects the plane $\lambda_2 = 0$ transversely. This concludes the proof of Theorem 5.10.

5.6 Nilpotent of codimension three in $\mathbb{R}^3$

In this section we consider C^∞ vector fields X defined in a neighborhood of $0 \in \mathbb{R}^3$ such that $X(0) = 0$ and $DX(0)$ is linearly conjugate to

$$x_2 \frac{\partial}{\partial x_1} + x_3 \frac{\partial}{\partial x_2}. \tag{5.33}$$

As explained in §5.4, X can be written as

$$x_2 \frac{\partial}{\partial x_1} + x_3 \frac{\partial}{\partial x_2} + f(x) \frac{\partial}{\partial x_3}$$

where $x = (x_1, x_2, x_3)$ and $f(x) = O(\|x\|^2)$. The condition $\frac{\partial^2 f}{\partial x_1^2}(0) \neq 0$ characterizes the 3-dimensional nilpotent singularity of codimension three. Any generic three-parameter unfolding of this singularity can be written as

$$x_2 \frac{\partial}{\partial x_1} + x_3 \frac{\partial}{\partial x_2} + \left(\mu_1 + \mu_2 x_2 + \mu_3 x_3 + x_1^2 + h(x, \mu)\right) \frac{\partial}{\partial x_3} \tag{5.34}$$

where $\mu = (\mu_1, \mu_2, \mu_3)$,

$$h(0, \mu) = 0, \quad \frac{\partial h}{\partial x_i}(0, \mu) = 0, \ \text{for } i = 1, 2, 3, \quad \frac{\partial^2 h}{\partial x_1^2}(0, \mu) = 0$$

and

$$h(x, \mu) = O(\|(x, \mu)\|^2), \quad h(x, \mu) = O(\|(x_2, x_3)\|).$$

Our main result in this section is the following:

Theorem 5.11. *There exists a curve* $B \subset \mathbb{R}^3$ *with an end at the origin, such that for each* $(\mu_1, \mu_2, \mu_3) \in B$, *system (5.34) exhibits a Bykov cycle.*

Blowing-up coordinates and parameters in (5.34) as done in (5.22)

$$\begin{aligned}
\mu_1 = \varepsilon^6 v_1, \quad \mu_2 = \varepsilon^2 v_2, \quad \mu_3 = \varepsilon v_2, \\
x_1 = \varepsilon^3 y_1, \quad x_2 = \varepsilon^4 y_2, \quad x_3 = \varepsilon^5 y_3,
\end{aligned} \tag{5.35}$$

we get, after rescaling time by a factor ε, the expression (5.23) for $n = 3$,

$$y_2 \frac{\partial}{\partial y_1} + y_3 \frac{\partial}{\partial y_2} + \left(v_1 + v_2 y_2 + v_3 y_3 + y_1^2 + \varepsilon \kappa y_1 y_2 + O(\varepsilon^2)\right) \frac{\partial}{\partial y_3} \tag{5.36}$$

where we assume that

$$\kappa = \frac{\partial^2 h}{\partial x_1 \partial x_2}(0, 0) \neq 0.$$

As usual, we consider $y = (y_1, y_2, y_3)$ varying in an arbitrarily big compact in $\mathbb{R}^3$ and $\nu = (\nu_1, \nu_2, \nu_3) \in \mathbb{S}^2$. When appropriate, we will also use directional blow-ups, by simply assuming that a convenient ν_i equals one in (5.35) and the other parameters vary in $\mathbb{R}^2$. Given this scenario, the first big challenge is to understand the dynamics exhibited by the limit family

$$y_2 \frac{\partial}{\partial y_1} + y_3 \frac{\partial}{\partial y_2} + \left(\nu_1 + \nu_2 y_2 + \nu_3 y_3 + y_1^2\right) \frac{\partial}{\partial y_3} \qquad (5.37)$$

obtained by taking $\varepsilon = 0$ in (5.36). Note that, as already discussed in §5.4, one only needs to consider $\nu_1 \leqslant 0$ and $\nu_3 \leqslant 0$.

Elementary bifurcations in the limit family were already studied in Dumortier, Ibáñez, and Kokubu (2001b). Note that there is a unique equilibrium point at $P_0 = (0, 0, 0)$ when $\nu_1 = 0$ and two equilibria

$$P_\pm = (\pm\sqrt{-\nu_1}, 0, 0)$$

when $\nu_1 < 0$. Figure 5.8 shows a partial bifurcation diagram.

When $(\nu_1, \nu_2, \nu_3) = (0, -1, 0)$ system (5.37) exhibits an equilibrium point at $(0, 0, 0)$ where the linear part has a pair of pure imaginary eigenvalues and the other equals 0. Equilibria with such eigenvalues are said Hopf–Zero singularities. Dynamics arising in the neighborhood of a Hopf–Zero singularity are discussed in Guckenheimer and Holmes (2002) and Kuznetsov (2004) and the arising of chaotic dynamics close to this singularity is explained in §5.8. There is a Bogdanov–Takens singularity at $(\nu_1, \nu_2, \nu_3) = (0, 0, -1)$. Details about the unfolding of the Bogdanov–Takens singularity are included in §5.5. Elementary Hopf and saddle-node bifurcations were described in §5.3.

As explained in Dumortier, Ibáñez, and Kokubu (2001b), when $\nu_2 \geqslant 0$ the dynamics is rather simple because there is a Lyapunov function, that is, there exists a scalar function defined on the phase space which is increasing along the orbits. It is proved that the system is gradient-like and hence, the maximal invariant set consists of the equilibrium points $P_\pm$ and heteroclinic orbits from P_- to P_+, which are saddles with stability indices 1 and 2, respectively. Existence of at least one heteroclinic connection is proved in Dumortier, Ibáñez, and Kokubu (2001b), but in spite of the intersection between the 2-dimensional invariant manifolds is likely to be unique, there is no a proof of this fact remains as an open problem.

We are particularly interested in the dynamics close to the reversibility curve

$$\mathcal{T} = \{(\nu_1, \nu_2, \nu_3) \in \mathbb{S}^2 : \nu_3 = 0\}.$$

As already explained, the dynamics when $\nu_2 \geqslant 0$ is simple, and we can pay attention only to the hemisphere with $\nu_2 < 0$. Under this assumption it is more

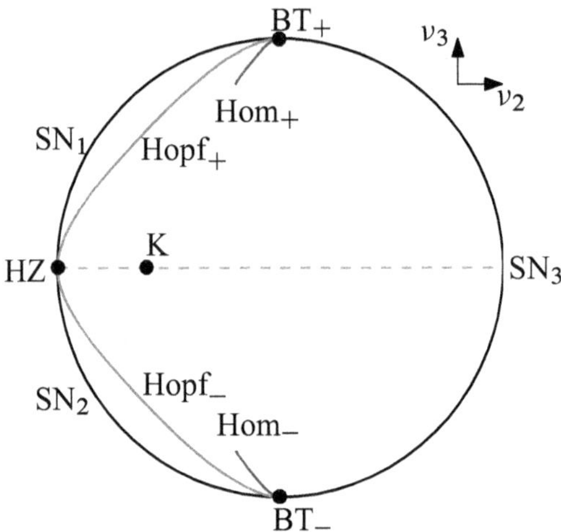

Figure 5.8: *Partial bifurcation diagram for the limit family (5.37). There are three local bifurcations of codimension two: a Hopf–Zero bifurcation (labelled HZ) when $(v_1, v_2, v_3) = (0, -1, 0)$ and two Bogdanov–Takens bifurcation (labelled BT^+ and BT^-) when $(v_1, v_2, v_3) = (0, 0, \pm 1)$. There are three lines of saddle-node bifurcation: $SN_1 = \{(v_1, v_2, v_3) \in \mathbb{S}^2 : v_1 = 0, -1 < v_2 < 0, v_3 < 0\}$, $SN_2 = \{(v_1, v_2, v_3) \in \mathbb{S}^2 : v_1 = 0, -1 < v_2 < 0, v_3 > 0\}$ and $SN_3 = \{(v_1, v_2, v_3) \in \mathbb{S}^2 : v_1 = 0, 0 < v_2 \geqslant 1, v_3 \leqslant 0\}$. Two Hopf bifurcation curves join HZ with BT^+ and BT^-, respectively. There are two curves of homoclinic bifurcation with end points at BT^+ and BT^-, respectively. The point K correspond to the parameter value where we prove the existence of a Bykov cycle.*

convenient to consider a directional rescaling of the parameters. We use the same formulas as in the spherical rescaling defined in (5.35), but we take $v_2 = -1$ and assume that $(v_1, v_3) = (\bar{v}_1, \bar{v}_3) \in \mathbb{R}^2$ to obtain

$$y_2 \frac{\partial}{\partial y_1} + y_3 \frac{\partial}{\partial y_2} + \left(\bar{v}_1 - y_2 + \bar{v}_3 y_3 + y_1^2 + \varepsilon \kappa y_1 y_2 + O(\varepsilon^2)\right) \frac{\partial}{\partial y_3}, \quad (5.38)$$

When $\bar{v}_3 = 0$ and $\bar{v}_1 \leqslant 0$, family (5.38) becomes the well known Michelson system Michelson (1986), the equation satisfied for the traveling waves of the Kuramoto–Shivashinsky equation in one-dimensional media, Kuramoto and Tsuzuki (1976):

$$u_t + u_{xxxx} + u_{xx} + \frac{1}{2} u_x^2 = 0,$$

where $t \geqslant 0$ and $x \in \mathbb{R}$. In order to use results from the literature we introduce new variables and parameters

$$\hat{y}_1 = -2y_1, \quad \hat{y}_2 = -2y_2, \quad \hat{y}_3 = -2y_3, \quad c = \sqrt{-2\bar{v}_1}$$

to get

$$\hat{y}_2 \frac{\partial}{\partial \hat{y}_1} + \hat{y}_3 \frac{\partial}{\partial \hat{y}_2} + \left(c^2 - \hat{y}_2 + \bar{v}_3 \hat{y}_3 + \frac{1}{2}\hat{y}_1^2 - 2\varepsilon \kappa \hat{y}_1 \hat{y}_2 + O(\varepsilon^2)\right) \frac{\partial}{\partial \hat{y}_3}.$$
$$(5.39)$$

Taking $\bar{v}_3 = 0$ and $\varepsilon = 0$ in the above expression we obtain the precise equation studied by Michelson in Michelson (1986):

$$\hat{y}_2 \frac{\partial}{\partial \hat{y}_1} + \hat{y}_3 \frac{\partial}{\partial \hat{y}_2} + \left(c^2 - \hat{y}_2 - \frac{1}{2}\hat{y}_1^2\right) \frac{\partial}{\partial \hat{y}_3}. \quad (5.40)$$

It is quite common to write (5.40) as a family of third order differential equations

$$x'''(t) + x'(t) = c^2 - \frac{1}{2}x^2,$$

with $x = \hat{y}_1$.

We already know that system (5.40) is reversible with respect to the involution

$$R : (\hat{y}_1, \hat{y}_2, \hat{y}_3) \to (-\hat{y}_1, \hat{y}_2, -\hat{y}_3). \quad (5.41)$$

Let $Q_\perp = (\pm\sqrt{2}c, 0, 0)$ be the equilibrium points of the system. The following are essential features of the Michelson system:

(M1) Q_+ and Q_- are equilibrium points of saddle-focus type. Eigenvalues are, respectively, $\{-\lambda, \rho \pm i\omega\}$ and $\{\lambda, -\rho \pm i\omega\}$, with $\lambda > 0$, $\rho > 0$ and $\omega \neq 0$. As divergence equals zero, $\rho = \frac{\lambda}{2}$. Note that in case that a homoclinic orbits exist, they satisfy the Shilnikov condition regarding the ratio of the local expansion versus contraction.

(M2) It follows from Kuramoto and Tsuzuki (1976) that for $c = c_K = \sqrt{\alpha}$, with $\alpha = 15\sqrt{11/19^3}$, (5.40) exhibits a heteroclinic orbit Γ_1 from Q_- to Q_+, that is, one of the branches of $W^u(Q_-)$ coincides with one of the branches of $W^s(Q_+)$. The explicit solution along this heteroclinic connection is given by

$$p(t) = (p_1(t), p_2(t), p_3(t)),$$

with

$$
\begin{aligned}
p_1(t) &= \alpha(-9\tanh \beta t + 11\tanh^3 \beta t), \\
p_2(t) &= \dot{p}_1(t), \\
p_3(t) &= \ddot{p}_1(t),
\end{aligned}
\tag{5.42}
$$

where $\beta = \sqrt{11/19}/2$. Note that p_1 and p_3 are odd functions whereas p_2 is an even function. The orbit of p is shown in Figure 5.9 (blue line) and graphs of p_i, with $i = 1, 2, 3$ are given in Figure 5.10 (left panel).

In the next section we will see how to prove that, when $c = c_k$, there also exists an orbit $\Gamma_2 \subset W^s(Q_-) \cap W^u(Q_+)$ along which the two-dimensional invariant manifolds intersect transversally, at least from a topological point of view (see Figure 5.9).

5.6.1   A Bykov cycle in the Michelson system

The main result in this subsection is the following:

Theorem 5.12. *The Michelson system* (5.40) *exhibits a Bykov cycle for the parameter value* $c = c_K$.

Bykov cycles are considered in §3. It follows from **(M2)** that there exists a heteroclinic connection Γ_1 from Q_- to Q_+. Here we prove that there exits a heteroclinic connection Γ_2 from Q_+ to Q_- (see Figure 5.9 and Figure 5.10).

The existence of Γ_2 for c large enough has been proved in Dumortier, Ibáñez, and Kokubu (2001b) and Michelson (1986). A result of existence is given in Jones, Troy, and MacGillivray (1992) for $c = 1$ and Lau (1992) provides numerical evidences of the existence and transversality of Γ for all $c > 0$.

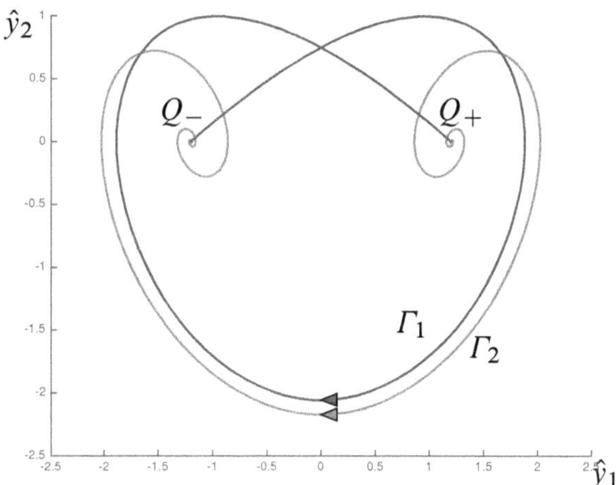

Figure 5.9: *Bykov cycle in the Michelson system. Red: Intersection between the two-dimensional invariant manifolds. Blue: Intersection between the one-dimensional invariant manifolds.*

However, there is no result valid for all $c > 0$. We will prove the existence and topological transversality when $c = c_K$ using techniques which are similar to those in Jones, Troy, and MacGillivray (1992), but also using the knowledge from Dumortier and Ibáñez (1996) about the behaviour of solutions of the Michelson system at infinity.

Remark 5.13. *An extensive use of computer assisted proofs allows to obtain a deep understanding of the Michelson systems for several values of c. See Kokubu, Wilczak, and Zgliczyński (2007) and Wilczak (2003, 2005) and references therein. See Lau (1992) for the notion of cocoon bifurcation linked to the Michelson system and also Dumortier, Ibáñez, and Kokubu (2006).*

Our first step is to understand how the unbounded orbits of system (5.40) escape to infinity. The behaviour of the Michelson system at infinity was studied in Dumortier, Ibáñez, and Kokubu (2001b). The proof of the result below can be found in Dumortier, Ibáñez, and Kokubu (2001b, Theorem 3).

Let $D = \{(x, y) \in \mathbb{R}^2 : x^2 + y^2 \leqslant 1\}$ and denote $C = D \times [0, 1]$.

Theorem 5.14. *Consider the limit family X_v given in (5.37). There exists a continuous map*

$$\Psi : C \times \mathbb{S}^2 \to A \subset \mathbb{R}^3$$

such that for all $v \in \mathbb{S}^2$, the domain $V_v = \Psi(C, v)$ is homeomorphic to C and:

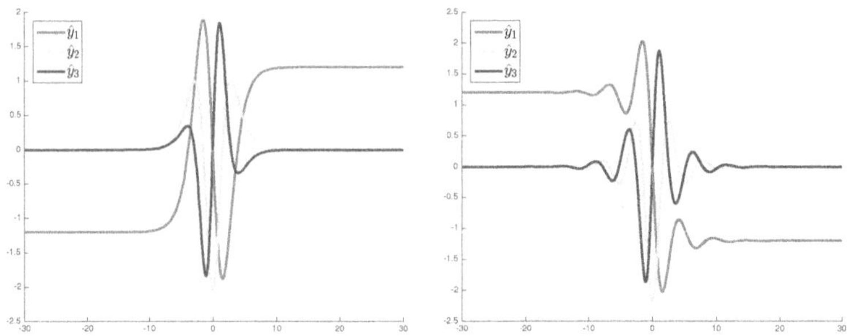

Figure 5.10: *Graphs of the solutions which parametrize the connections forming the Bykov cycle in the Michelson system. Left: Components of the solution corresponding to $W^s(Q_+) \cap W^u(Q_-)$ (the one-dimensional invariant manifolds). Right: Components of the solution corresponding to $W^s(Q_-) \cap W^u(Q_+)$ (the two-dimensional invariant manifolds). Note that $\hat{y}_1$ and $\hat{y}_3$ are odd functions and $\hat{y}_2$ is an even function.*

- $\Psi(\partial D \times [-1,1], v)$ *consists of regular orbits of* X_v,

- $\Psi(D \times \{-1,1\}, v)$ *is everywhere transverse to* X_v *which point inwards (resp. outwards)* V_v *on* $\Psi(D \times \{-1\}, v)$ *(resp. on* $\Psi(D \times \{1\}, v)$*).*

- *The maximal compact invariant set of* X_v *is contained in* V_v.

Following Dumortier, Ibáñez, and Kokubu (2001b) we refer to $\Psi(C \times [-1,1], v)$, $\Psi(D \times \{-1\}, v)$ and $\Psi(D \times \{1\}, v)$ as, respectively, the traffic regulator, the inset and the outset.

The phase-directional rescaling provides the behaviour at the infinity of vector fields in the family rescaling. The key of the proof of Theorem 5.14 is that there exists two equilibrium points at the infinity: a repeller $\widehat{Q}_+$ and an attractor $\widehat{Q}_-$. All unbounded forward (resp. backward) orbits tend to $\widehat{Q}_-$ (resp. $\widehat{Q}^+$) when time tends to ∞. Although it is not stated in Theorem 5.14, the inset $\Psi(D \times \{-1\}, v)$ (resp. the outset $\Psi(D \times \{1\}, v)$) is contained in a fundamental domain of the repeller (resp. attractor). This means that there are no orbits connecting the outset with the inset traveling outside the traffic regulator.

For future use we need to determine a fundamental domain of the attractor at the infinity. With that goal we consider the function

$$H(\hat{y}_1, \hat{y}_2, \hat{y}_3) = \hat{y}_3^2 + \hat{y}_2(\hat{y}_2 - 2c^2 + \hat{y}_1^2) \tag{5.43}$$

and observe that $\frac{dH}{dt} = 2\hat{y}_1\hat{y}_2^2$. Therefore H is strictly decreasing along orbits as long as they are contained in the region with $\hat{y}_1 < 0$ and $y_2 \neq 0$. Now, define the unbounded set

$$K = \{(\hat{y}_1, \hat{y}_2, \hat{y}_3) \in \mathbb{R}^3 : \hat{y}_1 < 0, \hat{y}_2 < 0, H(\hat{y}_1, \hat{y}_2, \hat{y}_3) < 0\}.$$

We can get a better understanding of the shape of K if we write

$$H(\hat{y}_1, \hat{y}_2, \hat{y}_3) = \left(\hat{y}_2 - \left(c^2 - \frac{\hat{y}_1^2}{2}\right)\right)^2 + \hat{y}_3^2 - \left(c^2 - \frac{\hat{y}_1^2}{2}\right)^2.$$

Given $h_0 > 0$, let us define

$$\varphi(\hat{y}_1) = \left(c^2 - \frac{\hat{y}_1^2}{2}\right)^2 - h_0.$$

Whenever $\varphi(\hat{y}_1) \geqslant 0$, for each $\hat{y}_1$ fixed, the set $H = -h_0$ is a circle with center at

$$C(\hat{y}_1) = \left(\hat{y}_1, \left(c^2 - \frac{\hat{y}_1^2}{2}\right), 0\right)$$

and radius $R(\hat{y}_1) = \sqrt{\varphi(\hat{y}_1)}$.

If $0 < h_0 \leqslant c^4$, $\varphi(\hat{y}_1) \geqslant 0$, when

$$\hat{y}_1 \in \left(-\infty, -\sqrt{2}\sqrt{c^2 + \sqrt{h_0}}\,\right]$$

$$\cup \left[-\sqrt{2}\sqrt{c^2 - \sqrt{h_0}}, \sqrt{2}\sqrt{c^2 - \sqrt{h_0}}\,\right]$$

$$\cup \left[\sqrt{2}\sqrt{c^2 + \sqrt{h_0}}, \infty\right),$$

and therefore the level surface $H = -h_0$ has three components. Note that $\hat{y}_2 \geqslant 0$ on the component correspondent to the middle interval. Therefore, the only of the three components which is contained in K is that correspondent to the left interval. On the other hand, if $c^4 < h_0$, $\varphi(\hat{y}_1) \geqslant 0$ when

$$\hat{y}_1 \in \left(-\infty, -\sqrt{2}\sqrt{c^2 + \sqrt{h_0}}\,\right) \cup \left(\sqrt{2}\sqrt{c^2 + \sqrt{h_0}}, \infty\right)$$

and the level surface has two components. It is straightforward to check that $\hat{y}_2 < 0$ on the component correspondent to the left interval. Hence such component is the only one contained in K. Finally, we conclude that K could be written as:

$$\{(\hat{y}_1, \hat{y}_2, \hat{y}_3) \in \mathbb{R}^3 : \hat{y}_1 < -\sqrt{2}c, H(\hat{y}_1, \hat{y}_2, \hat{y}_3) < 0\}.$$

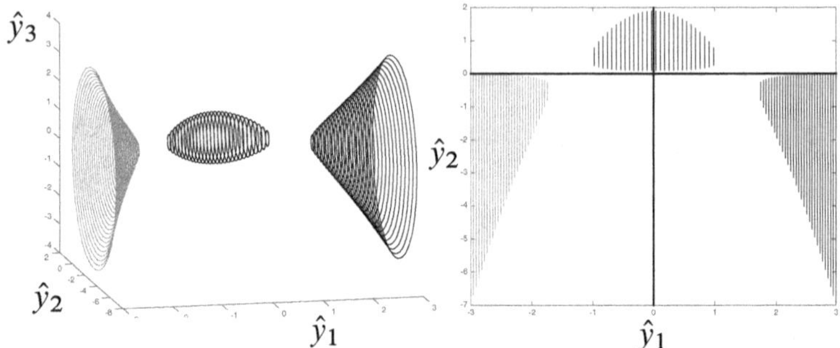

Figure 5.11: *Components of the level surface $H = -h_0$ for $0 < h_0 \leq c^4$.
The component in red colour belongs to the set K.*

Lemma 5.15. *The set K satisfies the following properties:*

1. *It is positively invariant.*

2. *All forward orbits of (5.40) contained in K are unbounded.*

3. *Let $\gamma = \{(\hat{y}_1(t), \hat{y}_2(t), \hat{y}_3(t)) : t \in I\}$ be an orbit of (5.40), where
   $I = (\rho_-, \rho_+)$ is the maximal interval of existence of the solutions. If γ is
   forward unbounded, there exists $\bar{t} > 0$ such that $(\hat{y}_1(t), \hat{y}_2(t), \hat{y}_3(t)) \in K$
   for all $t \geq \bar{t}$ and moreover*

$$\lim_{t \to \rho_+} \hat{y}_1(t) = \lim_{t \to \rho_+} \hat{y}_2(t) = \lim_{t \to \rho_+} \hat{y}_3(t) = -\infty.$$

4. *Let $\{(\hat{y}_1(t), \hat{y}_2(t), \hat{y}_3(t)) : t \in I\}$ be an orbit of (5.40) with $\hat{y}_1(0) =
   \hat{y}_3(0) = 0$. There exists $q > 0$ such that if $\hat{y}_2(0) < -q$ then $\hat{y}_2(t) < 0$
   for all $t \geq 0$ and there exists $\hat{t} > 0$ such that $(\hat{y}_1(t), \hat{y}_2(t), \hat{y}_3(t)) \in K$
   for all $t \geq \hat{t}$.*

Proof. Let $(p_1, p_2, p_3) \in K$ and define $h_0 = -H(p_1, p_2, p_3) < 0$. As already
argued, the component of $H = -h_0$ contained in K is a surface foliated by
circles which shrink to a point as $\hat{y}_1$ tends to $-\sqrt{2}\sqrt{c^2 + \sqrt{h_0}}$ from the left.
Since H decreases along the forward orbits on the whole surface, hence the
forward orbit of (p_1, p_2, p_3) is contained in

$$K_{h_0} = \{(\hat{y}_1, \hat{y}_2, \hat{y}_3) \in K : \hat{y}_1 < 0, \hat{y}_2 < 0, H(\hat{y}_1, \hat{y}_2, \hat{y}_3) \leq -h_0\}.$$

Therefore the set K is invariant by the forward flow.

On the other hand, suppose that the forward orbit of (p_1, p_2, p_3) is bounded,
hence its ω-limit is non-empty and must be contained in the set of points where

$\frac{dH}{dt}$ vanishes, but this set does not intersect K_{h_0}. It follows that all forward orbits with initial point in K are unbounded and K is contained in basin of attraction of $\widehat{Q}_-$, the above mentioned attractor at the infinity. In fact, as we prove below, K contains a fundamental domain of such attractor. This implies that any unbounded orbit of the Michelson system must cross K.

Take the following change of coordinates

$$\widehat{y}_1 = \frac{u}{\varepsilon^3} \qquad \widehat{y}_2 = \frac{v}{\varepsilon^4} \qquad \widehat{y}_1 = \frac{w}{\varepsilon^5} \tag{5.44}$$

where $\varepsilon \in (0, \infty)$ and $u^2 + v^2 + z^2 = 1$. Note that for each $(u, v, w) \in \mathbb{S}^2$ fixed, the above expression provides a curve $(\widehat{y}_1(\varepsilon), \widehat{y}_2(\varepsilon), \widehat{y}_3(\varepsilon))$ in $\mathbb{R}^3$ such that $\|(\widehat{y}_1(\varepsilon), \widehat{y}_2(\varepsilon), \widehat{y}_3(\varepsilon))\|$ tends to ∞ as ε tends to 0. Using (5.44) we obtain a vector field on $\mathbb{S}^2 \times (0, \infty)$ which we can extend to $\mathbb{S}^2 \times [0, \infty)$. Note that the restriction of such vector field to $\mathbb{S}^2 \times \{0\}$ corresponds to the "infinity" of the Michelson system.

Since we are not interested in the "whole infinity", but only in the neighborhood of the attractor, we can consider a convenient chart on the sphere $u^2 + v^2 + w^2 = 1$. Namely, we take $v = -1$ in (5.44) and assume $(u, w) \in \mathbb{R}^2$. After division by ε we get:

$$\dot{\varepsilon} = \frac{1}{4}\varepsilon w,$$

$$\dot{u} = -1 + \frac{3}{4}uw, \tag{5.45}$$

$$\dot{w} = -\frac{1}{2}u^2 + \frac{5}{4}w^2 + \varepsilon^2 + c^2\varepsilon^6.$$

As expected, the plane $\varepsilon = 0$ is invariant for the above system and gives the behaviour at the infinity of (5.40), at least when $\widehat{y}_2 < 0$ because of our choice $v = -1$. It is straightforward that there are only two equilibrium points

$$\widehat{Q}_{\pm} = \left(0, \pm\frac{\sqrt{6}}{3}10^{1/4}, \pm\frac{2\sqrt{6}}{3}10^{-1/4}\right)$$

and also that $\widehat{Q}_-$ is an attractor and $\widehat{Q}_+$ a repeller. In the new coordinates the function H is given by $H(\varepsilon, u, w) = \widehat{H}(\varepsilon, u, w)/\varepsilon^{10}$ with

$$\widehat{H}(\varepsilon, u, w) = w^2 - u^2 + \varepsilon^2 + 2c^2\varepsilon^6.$$

Moreover, the set K transforms into

$$\widehat{K} = \{(\varepsilon, u, w) \in \mathbb{R}^3 \ : \ \varepsilon > 0, (u/\varepsilon^3, -1/\varepsilon^4, w/\varepsilon^5) \in K\}$$

$$= \{(\varepsilon, u, w) \in \mathbb{R}^3 \ : \ \varepsilon > 0, u < 0, \widehat{H}(\varepsilon, u, w) < 0\}.$$

Since $\widehat{H}(\widehat{Q}_-) < 0$ and the second component of $\widehat{Q}_-$ is negative too, there exists $r > 0$ such that $B(\widehat{Q}_-, r) \cap \{(\varepsilon, u, v) \in \mathbb{R}^3 : \varepsilon > 0\} \subset \widehat{K}$. Given a fundamental domain D of the attractor contained in $B(\widehat{Q}_-, r)$, it follows that all orbits in $\{(\varepsilon, u, v) \in \mathbb{R}^3 : \varepsilon > 0\}$ tending to $\widehat{Q}_-$ as $t \to \infty$ must cut $D \cap \{(\varepsilon, u, v) \in \mathbb{R}^3 : \varepsilon > 0\} \subset \widehat{K}$ and therefore, as already said before, all the unbounded orbits of the Michelson system must enter in K. Taking into account that the attractor has been found in the chart with $v = -1$ and also that the u and w-components of the attracting equilibrium point are negative, it also follows that any unbounded forward orbit is such that all its components tend to $-\infty$.

Studying the phase portrait of the planar vector field obtained by taking $\varepsilon = 0$ in (5.45), it follows that $(0, 0, 0)$ belongs to the basin of attraction of $\widehat{Q}_-$. Hence, there exists $a > 0$ such that, for $\varepsilon \in (0, a)$, the forward orbit of $(\varepsilon, 0, 0)$ intersects $\widehat{K}$. Property (4) follows taking into account that the ε axis is sent to the negative $\hat{y}_2$-axis by the change (5.44) when $v = -1$. In particular we can take $q = -1/a^4$. $\qquad\qquad\qquad\qquad\qquad\qquad\qquad\qquad\qquad\qquad\qquad\quad$ $\square$

The following result is one of the main ingredients in the proof of the existence of a heteroclinic orbit connecting the equilibrium points along the two-dimensional invariant manifolds:

Lemma 5.16. *Let $(\hat{y}_1(t), \hat{y}_2(t), \hat{y}_3(t))$ be a solution of (5.40) for $c = c_K$ with $\hat{y}_1(0) = \hat{y}_2(0) = 0$ and $\hat{y}_3(0) \geq 0$. Then there exist values $a > 0$ and $b > a$ such that*

- $\hat{y}_2(t) > 0$ *for all $t \in (0, a)$,*

- $\hat{y}_2(a) = 0$,

- $\hat{y}_3(t) < 0$ *for all $t \in [a, b]$,*

- $\hat{y}_1(b) = 0$.

Lemma 5.16 is entirely technical and the proof is not provided here (see Ibáñez and Rodríguez (2005, Lemma 3.1)). The result is illustrated in Figure 5.12.

Let us define the set D given by all values $\gamma < 0$ such that the forward orbit of $(0, \gamma, 0)$ in the Michelson system (5.40) cuts the region with $\hat{y}_1 > 0$. We can state the following result (see Proposition 3.2 in Ibáñez and Rodríguez (2005)), which is valid for all $c > 0$, not only for $c = c_K$.

Proposition 5.17. *For any $c > 0$, the set D is non-empty, open and bounded from below. Moreover $\gamma_0 = \inf D$ does not belong to D.*

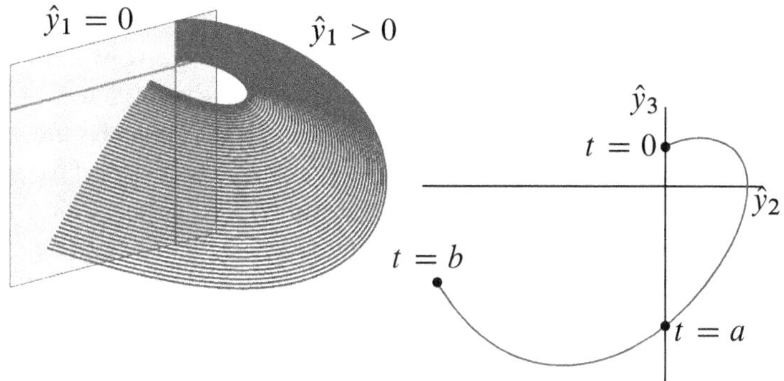

Figure 5.12: *Illustration of Lemma 5.16. Left panel: Forward orbits of the Michelson system with initial condition on the positive $\hat{y}_3$-axis as long they stay in the region $\hat{y}_1 \geq 0$. Right panel: Individual orbit projected on the plane with coordinates $(\hat{y}_2, \hat{y}_3)$.*

Proof. To prove that the set D is non-empty we first pay attention to the orbit of the origin. Let $(\hat{y}_1(t), \hat{y}_2(t), \hat{y}_3(t))$ be the solution of (5.40) satisfying $\hat{y}_1(0) = \hat{y}_2(0) = \hat{y}_3(0) = 0$. Since $\hat{y}_3'(0) = c^2 > 0$, there exists $\varepsilon > 0$ such that for any τ with $0 < \tau < c^2$, $\hat{y}_3'(t) > \tau$ for all $t \in (0, \varepsilon]$. It follows that, for all $t \in (0, \varepsilon]$,

$$\hat{y}_2'(t) = \hat{y}_3(t) > \tau t,$$

and therefore

$$\hat{y}_1'(t) = \hat{y}_2(t) > \tau \frac{t^2}{2},$$

and

$$\hat{y}_1(t) > \tau \frac{t^3}{6}.$$

Hence $\hat{y}_1(\varepsilon) > 0$. Because of the continuity with respect to the initial conditions, for $\delta > 0$ small enough, any solution $(\hat{y}_1^*(t), \hat{y}_2^*(t), \hat{y}_3^*(t))$ with $\hat{y}_1^*(0) = \hat{y}_3^*(t) = 0$ and $\hat{y}_2^*(t) = -\delta$ satisfies that $\hat{y}_1^*(\varepsilon) > 0$ and we conclude that the set D is non-empty. To prove that D is open an argument of continuity is also enough. Indeed, if a point of the $\hat{y}_2$-axis is such that its forward orbit cuts the region $\hat{y}_1 > 0$, all orbits starting close enough also cut such region.

To see that D is bounded below, we observe that, as stated in Lemma 5.15 (see item 4), there exists $q > 0$ such that all forward orbits starting on a point $(0, y_0, 0)$ with $y_0 < -q$ are contained on the region with $\hat{y}_2 < 0$. Since $\hat{y}_1' = \hat{y}_2$, it follows that $\hat{y}_1 < 0$ along such forward orbit. Hence $-q$ is a lower bound of D and there exists $\gamma_0 = \inf D$. Note that $\gamma_0 \notin D$ because D is open. $\qquad\square$

From now on, $U(t) = (\bar{y}_1(t), \bar{y}_1(t), \bar{y}_1(t))$ will be the solution of (5.40) with initial condition $U(0) = (0, \gamma_0, 0)$. We will show that, al least when $c = c_K$, the orbit of $(0, \gamma_0, 0)$ is a heteroclinic connection given as the intersection of the two-dimensional invariant manifolds. The argument is valid for all c for which a result as that in Lemma 5.16 is valid. Note that, in particular, we will find a heteroclinic connection at the initial condition on the $\hat{y}_2$-axis determined by $\inf D$ for all c close enough to c_K.

Theorem 5.18. *If $c = c_K$ then $\lim_{t \to \pm\infty} U(t) = Q_\mp$.*

Proof. First we will prove that $\bar{y}_1(t) < 0$ for all $t > 0$. Since $\gamma_0 \notin D$ (see Proposition 5.17), it follows that $\bar{y}_1(t) \leqslant 0$ for all $t > 0$. Assume that there exists $\bar{t} > 0$ such that $\bar{y}_1(\bar{t}) = 0$. Since $\bar{y}_1' = \bar{y}_2$, there must be $\bar{y}_2(\bar{t}) = 0$. Otherwise, if $\bar{y}_2(\bar{t}) > 0$, $\bar{y}_1(t)$ would be strictly positive for $t > \bar{t}$ close enough to $\bar{t}$ and if $\bar{y}_2(\bar{t}) < 0$, $\bar{y}_1(t)$ would be strictly positive for $t < \bar{t}$ close enough to $\bar{t}$, in either case we get a contradiction with the fact that $\bar{y}_1(t) \leqslant 0$ for all $t > 0$. Even more, $\bar{y}_3(\bar{t}) < 0$ since otherwise, as follows from Lemma 5.16 $\bar{y}_1(t)$ would be strictly positive for $t > \bar{t}$ close to $\bar{t}$. Therefore we have an orbit connecting the negative $\bar{y}_2$-axis with the negative $\bar{y}_3$-axis. Its image by the reversibility map R given in (5.41) is and orbit connecting the positive $\bar{y}_3$-axis with the negative $\bar{y}_2$-axis. This contradicts Lemma 5.16.

Assume now that the orbit of $(0, \gamma_0, 0)$ is forward unbounded. It follows from Lemma 5.15 that there exists $\bar{t} > 0$ such that $U(t) \in K$ for all $t \geqslant \bar{t}$. By the continuity of the flow with respect to the initial conditions, there exists $\gamma > \gamma_0$ such that the forward orbit $(\bar{y}_1^*(t), \bar{y}_2^*(t), \bar{y}_3^*(t))$ of $(0, \gamma, 0)$ also enters K and stays there for all $t \geqslant \bar{t}$ while it is also satisfied that $\bar{y}_1^*(t) < 0$ for all $t \in (0, \bar{t}]$. Therefore $\gamma > \gamma_0$ is a lower bound of D and we have a contradiction. Note that, because of the reversibility, if the forward orbit of a point in the $\hat{y}_2$ axis is forward bounded, the backward orbit is also bounded.

Provided that $U(t)$ is a bounded solution, the ω-limit L of the orbit of $(0, \gamma_0, 0)$ is non-empty. On the other hand, because $\bar{y}_1 < 0$ along the orbit, the function H as defined in (5.43) is decreasing and hence L must be contained in the set

$$C = \{(\hat{y}_1, \hat{y}_2, \hat{y}_3) \in \mathbb{R}^3 \, : \, \hat{y}_1 \leqslant 0, H(\hat{y}_1, \hat{y}_2, \hat{y}_3) \leqslant \gamma_0(\gamma_0 - c^2)\}.$$

Moreover L must be contained in the set of points where $\frac{dH}{dt} = 0$. Therefore L is either contained in the plane $\hat{y}_1 = 0$ or in the plane $\hat{y}_2 = 0$ with $\hat{y}_1 < 0$. The only invariant set contained in that union of planes is the equilibrium point Q_-. Taking into account the reversibility, we get the required result. $\qquad\square$

To conclude with our study of Γ_2, we will prove that the intersection between the two dimensional invariant manifolds along Γ_2 is locally unique and topologically transversal.

Given $\varepsilon > 0$, let us define

$$D = \{(\hat{y}_1, \hat{y}_2, \hat{y}_3) \in \mathbb{R}^3 \,:\, \hat{y}_1 = 0, \sqrt{(\hat{y}_1 - \gamma_0)^2 + \hat{y}_3^2} < \varepsilon\}.$$

and also Π^u and Π^s as the connected components of $W^u(Q_+) \cap \{\hat{y}_1 = 0\}$ and $W^s(Q_-) \cap \{\hat{y}_1 = 0\}$, respectively, which contain the point $(0, \gamma_0, 0)$. Note that if ε is small enough, both, Π^u and Π^s split the disc D into two connected components. We say that the intersection between the two dimensional invariant manifolds is locally unique if $\Pi^s \cap \Pi^u = \{(0, \gamma_0, 0)\}$. We say that the intersection is topologically transversal if the two components of $\Pi^u \setminus \{(0, \gamma_0, 0)\}$ (resp. $\Pi^s \setminus \{(0, \gamma_0, 0)\}$) are not contained in the same component of $D \setminus \Pi^s$ (resp. $D \setminus \Pi^u$).

Theorem 5.19. *The intersection of the two-dimensional invariant manifolds along Γ is locally unique and topologically transversal.*

Proof. Note that due to the reversibility, we only need to show that $\Pi^s \cap \{\hat{y}_3 = 0\} = \{(0, \gamma_0, 0)\}$ and also that the two segments given by $(D \cap \{\hat{y}_3 = 0\}) \setminus \{p_0\}$ belong to different components of $D \setminus \Pi^s$.

Since the vector field is analytic, invariant manifolds are also locally analytic. Hence one can write Π^s as the graph of an analytic function, either defined for $\hat{y}_2$ in a neighborhood of γ_0 or for $\hat{y}_3$ defined in a neighborhood of 0. In any case, the local uniqueness follows because the zeroes of an analytic function are isolated.

Now, notice that $W^u(Q_-)$ splits into two branches. One of them corresponds to the heteroclinic connection with Q_+ and intersects the plane $\hat{y}_1 = 0$. A local linear analysis at Q_- shows that the other branch is contained in K and hence $\hat{y}_1 < 0$ along the orbit. As follows from our study, if ε is small enough, orbits starting at one of the components $D \setminus \Pi^s$ will intersect $\hat{y}_1 = 0$ following the bounded branch of $W^u(Q_-)$. We call D_1 to this component. On the other hand, orbits starting at the other component will enter in K without crossing again the plane $\hat{y}_1 - 0$. Denote by D_2 this component.

Assume that $(D \cap \{\hat{y}_3 = 0\}) \setminus \{p_0\} \subset D_1$. Hence there exist points $(0, \gamma, 0) \in D$ with $\gamma < \inf D$. Otherwise, if we assume that $(D \cap \{\hat{y}_3 = 0\}) \setminus \{p_0\} \subset D_2$ then there exist lower bounds of D which are larger than $\inf D$. In either case we get a contradiction. $\qquad\square$

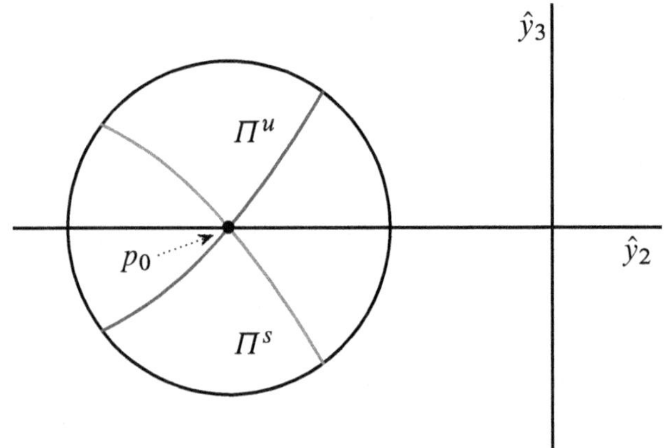

Figure 5.13: *Illustration of the transversality and local uniqueness. $\Pi^s = W^s(Q_-) \cap B_\varepsilon$ and $\Pi^u = W^u(Q_+) \cap B_\varepsilon$. The point $p_0 = (0, \gamma_0, 0)$ corresponds to the intersection of Γ with $\hat{y}_1 = 0$.*

5.6.2　Bykov cycles in the unfolding

To prove Theorem 5.11 we have to study the splitting function determined when moving parameters the connection along the one-dimensional invariant manifolds breaks. With that goal in mind, we consider the family (5.39) written as

$$\dot{x} = f(x) + g(\lambda, x), \tag{5.46}$$

where

$$\lambda = (\lambda_1, \lambda_2, \lambda_3) = \left(c^2 - c_K^2, \bar{v}_3, \varepsilon\right),$$

varies in a neighborhood of $(0, 0, 0) \in \mathbb{R}^3$,

$$f(x_1, x_2, x_3) = \left(x_2, x_3, c_k^2 - x_2 - \frac{1}{2}x_1^2\right)$$

and

$$g(x_1, x_2, x_3) = (0, 0, \lambda_1 + \lambda_2 x_3 - 2\lambda_3 \kappa x_1 x_2 + O(\lambda_3^2)).$$

Bifurcations exhibited when $\lambda_3 > 0$ correspond to bifurcations observed in any generic unfolding of the three-dimensional nilpotent singularity of codimension three.

We will prove that for parameter values along a curve arising from the origin, there are heteroclinic orbits connecting the equilibrium points along the one-dimensional invariant manifolds, that is, the unfolding of the heteroclinic connection Γ_1 inside family (5.46) is generic.

Since $\dim\left(T_P W^u(P) \cap T_P W^s(P)\right) = 1$, where $P \in \gamma_1$ is an arbitrary point, the heteroclinic orbit Γ_1 is non-generic of codimension two in the sense of Definition A.22. Hence (see Proposition A.15 and Proposition A.21), the variational equation

$$\begin{pmatrix} \dot{z}_1 \\ \dot{z}_2 \\ \dot{z}_3 \end{pmatrix} = Df(p(t)) \begin{pmatrix} z_1 \\ z_2 \\ z_3 \end{pmatrix} = \begin{pmatrix} 0 & 1 & 0 \\ 0 & 0 & 1 \\ -p_1(t) & -1 & 0 \end{pmatrix} \begin{pmatrix} z_1 \\ z_2 \\ z_3 \end{pmatrix}$$

with $p(t) = (p_1(t), p_2(t), p_3(t))$ as given in (5.42), has a unique (up to multiplicative constants) bounded solution $f(p(t))$. Moreover, the adjoint variational equation

$$\begin{pmatrix} \dot{w}_1 \\ \dot{w}_2 \\ \dot{w}_3 \end{pmatrix} = -Df(p(t))^* \begin{pmatrix} w_1 \\ w_2 \\ w_3 \end{pmatrix} = \begin{pmatrix} 0 & 0 & p_1(t) \\ -1 & 0 & 1 \\ 0 & -1 & 0 \end{pmatrix} \begin{pmatrix} w_1 \\ w_2 \\ w_3 \end{pmatrix}$$
$$\tag{5.47}$$

has a pair of linearly independent bounded solutions.

Let $\varphi(t) = (\varphi_1(t), \varphi_2(t), \varphi_3(t))$ and $\psi(t) = (\psi_1(t), \psi_2(t), \psi_3(t))$ be two linearly independent bounded solutions of the adjoint variational equation. Since $\varphi(t) \wedge \psi(t)$ is a bounded solution of the variational equation, it follows that the plane determined by $\varphi(t)$ and $\psi(t)$ is orthogonal to $f(p(t))$ for all $t \in \mathbb{R}$. Therefore all solutions of the adjoint variational equation with initial conditions on $f(p(0))^{\perp}$ are bounded solutions.

Note that the adjoint variational equation is time-reversible with respect to the involutions:

$$R_1 : (w_1, w_2, w_3) \to (-w_1, w_2, -w_3)$$

and

$$R_2 : (w_1, w_2, w_3) \to (w_1, -w_2, w_3)$$

On the other hand, $p(0) = (0, -9\alpha\beta, 0)$ and hence $f(p(0)) = (-9\alpha\beta, 0, c_K^2 + 9\alpha\beta)$. It easily follows that

$$f(p(0))^{\perp} = \left\langle (0, -1, 0), \left(1, 0, 1 + \frac{c_K^2}{9\alpha\beta} \right) \right\rangle.$$

Let φ and ψ the linearly independent bounded solutions of the adjoint variational equations with $\varphi(0) = (0, -1, 0)$ and $\psi(0) = (1 + c_K^2/9\alpha\beta, 0, 1)$. Our choice is such φ and ψ satisfy nice parity properties. Namely, $\varphi(0)$ belongs to the fixed points for the involution R_1 and hence φ_1 and φ_3 are odd function, whereas φ_2 is an even function. Similarly, $\psi(0)$ belongs to the fixed points for the involution R_2 and hence ψ_2 is an odd functions, whereas ψ_1 and ψ_3 are even functions.

To study the persistence of the heteroclinic orbit Γ_1 we consider the splitting function $\xi^\infty(\lambda)$ which provides the displacement between $W^u(Q_-^\lambda)$ and $W^s(Q_+^\lambda)$ on $f(p(0))^\perp$, where Q_-^λ and Q_+^λ are the continuation with respect to λ of the equilibrium points Q_- and Q_+, respectively. Note that

$$\xi^\infty = (\xi_1^\infty, \xi_2^\infty) : \Lambda \subset \mathbb{R}^3 \to \mathbb{R}^2$$

and $\xi^\infty(\lambda) = 0$ means that there are branches of the one-dimensional invariant manifolds which are coincident.

Let

$$D_\lambda \xi^\infty(0) = \begin{pmatrix} \xi_{1,\lambda_1}^\infty(0) & \xi_{1,\lambda_2}^\infty(0) & \xi_{1,\lambda_3}^\infty(0) \\ \xi_{2,\lambda_1}^\infty(0) & \xi_{2,\lambda_2}^\infty(0) & \xi_{2,\lambda_3}^\infty(0) \end{pmatrix}$$

The bifurcation equation $\xi^\infty(\lambda) = 0$ can be studied by means of the Implicit Function Theorem. Namely, if $\mathrm{rank} D_\lambda \xi^\infty(0) = 2$ the system has heteroclinic orbits connecting the equilibrium points along the one-dimensional invariant manifolds for parameter values along a curve $\mathscr{H}$ through $\lambda = 0$ where the tangent subspace is given by the intersection of the planes:

$$\xi_{1,\lambda_1}^\infty(0)\lambda_1 + \xi_{1,\lambda_2}^\infty(0)\lambda_2 + \xi_{1,\lambda_3}^\infty(0)\lambda_3 = 0$$
$$\xi_{2,\lambda_1}^\infty(0)\lambda_1 + \xi_{2,\lambda_2}^\infty(0)\lambda_2 + \xi_{2,\lambda_3}^\infty(0)\lambda_3 = 0.$$

According to Theorem B.2

$$\xi_{1,\lambda_1}^\infty(0) = \int_{-\infty}^{\infty} \langle \varphi(s), D_{\lambda_1} g(0, p(s)) \rangle \, ds = \int_{-\infty}^{\infty} \varphi_3(s) \, ds$$

$$\xi_{1,\lambda_2}^\infty(0) = \int_{-\infty}^{\infty} \langle \varphi(s), D_{\lambda_2} g(0, p(s)) \rangle \, ds = \int_{-\infty}^{\infty} \varphi_3(s) p_3(s) \, ds$$

$$\xi_{1,\lambda_3}^\infty(0) = \int_{-\infty}^{\infty} \langle \varphi(s), D_{\lambda_3} g(0, p(s)) \rangle \, ds = \int_{-\infty}^{\infty} -2\kappa \varphi_3(s) p_1(s) p_2(s) \, ds$$

$$\xi_{2,\lambda_1}^\infty(0) = \int_{-\infty}^{\infty} \langle \psi(s), D_{\lambda_1} g(0, p(s)) \rangle \, ds = \int_{-\infty}^{\infty} \psi_3(s) \, ds$$

$$\xi_{2,\lambda_2}^\infty(0) = \int_{-\infty}^{\infty} \langle \psi(s), D_{\lambda_2} g(0, p(s)) \rangle \, ds = \int_{-\infty}^{\infty} \psi_3(s) p_3(s) \, ds$$

$$\xi_{3,\lambda_3}^\infty(0) = \int_{-\infty}^{\infty} \langle \psi(s), D_{\lambda_3} g(0, p(s)) \rangle \, ds = \int_{-\infty}^{\infty} -2\kappa \psi_3(s) p_1(s) p_2(s) \, ds.$$

From the parities of the components of p, φ and ψ, it follows that

$$\xi_{1,\lambda_1}^{\infty}(0) = \xi_{2,\lambda_2}^{\infty}(0) = \xi_{2,\lambda_3}^{\infty}(0) = 0.$$

Later we will argue that

$$\xi_{1,\lambda_2}^{\infty}(0) \neq 0 \qquad \xi_{1,\lambda_3}^{\infty}(0) \neq 0 \qquad \xi_{2,\lambda_1}^{\infty}(0) \neq 0. \tag{5.48}$$

As already mentioned, by the Implicit Function Theorem, there exists a bifurcation curve $\mathcal{H} \subset \Lambda$ corresponding to parameter values for which the system exhibits a heteroclinic connection along the one-dimensional invariant manifolds. Moreover, the vector

$$\left(0, -\frac{-\xi_{1,\lambda_3}}{\xi_{1,\lambda_2}}, 1 \right)$$

is tangent to $\mathcal{H}$ at $\lambda = 0$. Therefore, $\mathcal{H}$ intersects the plane $\lambda_3 = 0$ and so, this bifurcation curve is observed in any generic unfolding of the three-dimensional nilpotent singularity of codimension three. On the other hand, since Γ_2 is a topologically transverse intersection between the two-dimensional invariant manifolds, it is persistent for λ nearby enough to $\lambda = 0$. Thus, we conclude that $\mathcal{H}$ is not only a bifurcation curve for heteroclinic orbits along the one-dimensional invariant manifolds, but also a bifurcation curve corresponding to parameter values for which the system exhibits a topological Bykov cycle. So, we have proved Theorem 5.11. It follows from the bifurcation theorem that fixing λ_3 the family (5.39) is a generic unfolding a Bykov cycle. As argued in Chapter 3, any generic unfolding of a Bykov cycle contains a spiral-shaped Shilnikov bifurcation curve. Hence family (5.39) exhibits a spiraling-sheet bifurcation surface $\mathcal{H}_+$ corresponding to parameter values for which the system has a Shilnikov homoclinic orbit to P_+^{λ}. Note that the Shilnikov condition is open and hence it follows from property **(M1)**. Moreover, the trace of the linear part at P_+^{λ} is given by λ_2 and hence the dissipative condition is also satisfied in $\mathcal{H}_+ \cap \{\lambda \in \mathbb{R}^3 : \lambda_2 < 0\}$.

Remark 5.20. *Note that there also exists a spiraling-sheet bifurcation surface $\mathcal{H}_-$ corresponding to parameter values for which the system has a Shilnikov homoclinic orbit to P_-^{λ}. Coexistence of strange attractors and repellers is plausible.*

All we need to do to conclude the whole argumentation is to prove that indeed inequalities in (5.48) are satisfied. Taking into account the parity properties of the bounded solutions φ and ψ, one only needs to prove that:

$$\xi^{\infty}_{1,\lambda_2}(0) = 2 \int_0^{\infty} \varphi_3(s)p_3(s)\,ds \neq 0$$

$$\xi^{\infty}_{1,\lambda_3}(0) = -4\kappa \int_0^{\infty} \varphi_3(s)p_1(s)p_2(s)\,ds \neq 0$$

$$\xi^{\infty}_{2,\lambda_1}(0) = 2 \int_0^{\infty} \psi_3(s)\,ds \neq 0.$$

Each bounded solution of the adjoint variational equation (5.47) satisfies an orthogonality condition with respect to $f(p(t))$. Namely

$$p_2(t)w_1(t) + p_3(t)w_2(t) + (c_K^2 - p_2(t) - \frac{(p_1(t))^2}{2})w_3(t) = 0.$$

Taking into account such condition we can write (5.47) as

$$\dot{w}_1(t) = -\frac{1}{p_2(t)}\left(p_3(t)w_2(t) + (c_K^2 - p_2(t) - \frac{(p_1(t))^2}{2})w_3(t)\right)$$

$$\dot{w}_2(t) = \frac{1}{p_2(t)}\left(p_3(t)w_2(t) + \left(c_K^2 - \frac{(p_1(t))^2}{2}\right)w_3(t)\right)$$

$$\dot{w}_3(t) = -w_2(t).$$

The last two equations in the above system are decoupled and we can write them as:

$$A(t)\widetilde{w}'(t) = B(t)\widetilde{w}(t) \tag{5.49}$$

with $\widetilde{w}(t) = (w_2(t), w_3(t))$ and

$$A(t) = \begin{pmatrix} p_2(t) & 0 \\ 0 & 1 \end{pmatrix} \qquad B(t) = \begin{pmatrix} p_3(t) & c_K^2 - \dfrac{(p_1(t))^2}{2} \\ 0 & -1 \end{pmatrix}.$$

It easily follows that $p_2(t)$ vanishes at $t = t^*_{\pm} = \pm(1/\beta)\tanh^{-1}(\sqrt{3/11})$. Hence $A(t)$ is singular and equation (5.49) must treated as an algebraic differential equation. Numerical integration is required to prove that inequalities (5.48) are satisfied. We do not provide the details about this computation; they can be found in Barrientos, Ibáñez, and Rodríguez (2011). One has to split the integrals in $[0, \infty)$ in two, one in $[0, t_0]$ and the other in $[t_0, \infty)$ in such a way that $t^*_+ \in [0, t_0]$ and the integrals in $[t_0, \infty)$ are small. Integrals in $[0, t_0]$ can be approached combining a numerical method to solve (5.49) with a numerical method to compute the integrals. Matlab, for instance, provide numerical methods to solve algebraic differential equations. Hence, the value of the integrals in the finite interval $[0, t_0]$ can be approached with an error as small as required. On the other hand, the integrals in $[t_0, \infty)$ can be made as small as needed by choosing t_0 large enough. Hence the result follows.

5.7   Nilpotent of codimension four in $\mathbb{R}^4$

Consider C^∞ vector fields X defined in a neighborhood of $0 \in \mathbb{R}^4$ such that $X(0) = 0$ and $DX(0)$ is linearly conjugate to

$$x_2 \frac{\partial}{\partial x_1} + x_3 \frac{\partial}{\partial x_2} + x_4 \frac{\partial}{\partial x_3}. \tag{5.50}$$

As explained in §5.4, X can be written as

$$x_2 \frac{\partial}{\partial x_1} + x_3 \frac{\partial}{\partial x_2} + x_4 \frac{\partial}{\partial x_3} + f(x) \frac{\partial}{\partial x_4}$$

where $x = (x_1, x_2, x_3, x_4)$ and $f(x) = O(\|x\|^2)$. The condition $\frac{\partial^2 f}{\partial x_1^2}(0) \neq 0$ characterizes the four-dimensional nilpotent singularity of codimension four. Any generic four-parameter unfolding of this singularity can be written as

$$\begin{aligned} &x_2 \frac{\partial}{\partial x_1} + x_3 \frac{\partial}{\partial x_2} + x_4 \frac{\partial}{\partial x_3} \\ &+ \left(\mu_1 + \mu_2 x_2 + \mu_2 x_3 + \mu_4 x_4 + x_1^2 + h(x, \mu)\right) \frac{\partial}{\partial x_4} \end{aligned} \tag{5.51}$$

where $\mu = (\mu_1, \mu_2, \mu_3, \mu_4)$,

$$h(0, \mu) = 0, \quad \frac{\partial h}{\partial x_i}(0, \mu) = 0, \ \text{for } i = 1, \ldots, 4, \quad \frac{\partial^2 h}{\partial x_1^2}(0, \mu) = 0,$$

and

$$h(x, \mu) = O(\|(x, \mu)\|^2), \quad h(x, \mu) = O(\|(x_2, x_3, x_4)\|).$$

Our main result in this section is the following:

Theorem 5.21. *There exists a hypersurface $\mathscr{FF}$ in the parameter space such that for each $(\mu_1, \mu_2, \mu_3, \mu_4) \subset \mathscr{FF}$, system (5.51) exhibits a bifocus homoclinic orbit.*

Blowing-up coordinates and parameters in (5.51) as done in (5.22)

$$\mu_1 = \varepsilon^8 v_1, \quad \mu_2 = \varepsilon^3 v_2, \quad \mu_3 = \varepsilon^2 v_3, \quad \mu_4 = \varepsilon v_4,$$
$$x_1 = \varepsilon^4 y_1, \quad x_2 = \varepsilon^5 y_2, \quad x_3 = \varepsilon^6 y_3, \quad x_4 = \varepsilon^7 y_4, \tag{5.52}$$

we get, after rescaling time by a factor ε, the expression (5.23) for $n = 4$,

$$y_2 \frac{\partial}{\partial y_1} + y_3 \frac{\partial}{\partial y_2} + y_4 \frac{\partial}{\partial y_3}$$
$$+ \left(v_1 + v_2 y_2 + v_3 y_3 + v_4 y_4 + y_1^2 + \varepsilon \kappa y_1 y_2 + O(\varepsilon^2) \right) \frac{\partial}{\partial y_4} \tag{5.53}$$

where we assume that

$$\kappa = \frac{\partial^2 h}{\partial x_1 \partial x_2}(0, 0) \neq 0.$$

As usual, we consider $y = (y_1, y_2, y_3, y_4)$ varying in an arbitrarily big compact in $\mathbb{R}^4$ and $v \in \mathbb{S}^3$. The first big challenge is to understand the dynamics exhibited by the limit family

$$y_2 \frac{\partial}{\partial y_1} + y_3 \frac{\partial}{\partial y_2} + y_4 \frac{\partial}{\partial y_3} + \left(v_1 + v_2 y_2 + v_3 y_3 + v_4 y_4 + y_1^2 \right) \frac{\partial}{\partial y_4} \tag{5.54}$$

obtained by taking $\varepsilon = 0$ in (5.53). Note that, as already explained in §5.4, one only needs to consider $v_1 \leqslant 0$ and $v_4 \leqslant 0$.

Local bifurcations arising in the limit family are discussed in Drubi (2009). There is a unique equilibrium point at $P_0 = (0, 0, 0, 0)$ when $v_1 = 0$ and two equilibria $P_\pm = (\pm\sqrt{-v_1}, 0, 0, 0)$ when $v_1 < 0$. As expected, we find surfaces of saddle-node bifurcation and also surfaces of Hopf bifurcation. Regarding codimension two local bifurcations, there appear several cases of Hopf–Zero, Hopf–Hopf and Bogdanov–Takens bifurcations. There also appear two codimension three local bifurcations: a three-dimensional nilpotent singularity of codimension three and a Hopf–Bogdanov–Takens singularity. Most of these singulrities been already mentioned in different contexts along this chapter. Only the Hopf–Hopf and the Hopf–Bogdanov–Takens singularities appear by the first time in our discussion. Dynamics arising in the neighborhood of a Hopf–Hopf singularity are discussed in Guckenheimer and Holmes (2002) and Kuznetsov (2004) and the arising of chaotic dynamics close to this singularity is explained in §5.8. Hopf–Bogdanov–Takens singularities are four-dimensional of codimension three characterized by a linear part which is linearly conjugate to

$$y \frac{\partial}{\partial x} - v \frac{\partial}{\partial u} + u \frac{\partial}{\partial v}.$$

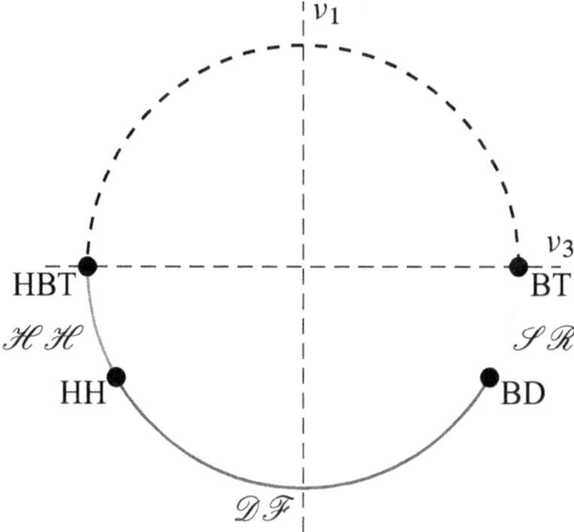

Figure 5.14: *Partial bifurcation diagram of the limit family (5.54) restricted to the reversibility curve.*

These singularities have not yet been classified. The only contribution in this regard is Drubi, Ibáñez, and Rivela (2019a), where a formal classification is provided. We say formal in the sense that topological types are determined for a truncated normal form, but the effect of higher order terms is not yet understood. Numerical evidences of the existence of chaotic dynamics in the unfolding of Hopf–Bogdanov–Takens singularities are provided in Drubi, Ibáñez, and Rivela (2019a).

For parameter values on the reversibility curve

$$\mathscr{T} = \{(\nu_1, \nu_2, \nu_3, \nu_4) \in \mathbb{S}^3 : \nu_2 = \nu_4 = 0\}$$

local bifurcations are depicted in Figure 5.14

The characteristic equations at the equilibrium points $P_\pm$ are given by

$$r^4 - \nu_3 r^2 \mp 2\sqrt{-\nu_1} = 0.$$

It is straightforward that P^+ is always hyperbolic for parameter values on $\mathscr{T}$. Nevertheless, the local behaviour in a neighborhood of P_- is much richer. Local bifurcations at P_- are depicted in Figure 5.14. It easily follows that:

- When $(\nu_1, \nu_2, \nu_3, \nu_4) = (0, 0, 1, 0)$ the linear part at P_- has a double zero eigenvalue. The other two are $\mathbf{i} 1$ and -1. At this point there is a Bogdanov–Takens bifurcation (see BT in Figure 5.14).

- When $(v_1, v_2, v_3, v_4) = (0, 0, -1, 0)$ the linear part at P_- has a double zero eigenvalue and a pair of pure imaginary eigenvalues. At this point we find a Hopf–Bogdanov–Takens bifurcation (see HBT in Figure 5.14).

- When $(v_1, v_2, v_3, v_4) = (\bar{v}_1, 0, \bar{v}_3, 0)$, with $v_3^2 = 8\sqrt{-v_1}$ and $v_3 > 0$, the linear part at P_- has a couple of double real eigenvalues, (see BD in Figure 5.14).

- When $(v_1, v_2, v_3, v_4) = (\bar{v}_1, 0, \bar{v}_3, 0)$, with $v_3^2 = 8\sqrt{-v_1}$ and $v_3 < 0$, the linear part at P_- has a couple of double pure imaginary eigenvalues, (see HH in Figure 5.14). This a degenerated Hopf–Hopf bifurcation.

- For parameter values in between BT and BD, the linear part has four non-zero real eigenvalues (see the arc denoted by $\mathscr{S}\mathscr{R}$ in Figure 5.14).

- For parameter values in between BD and HH, the linear part has four complex eigenvalues $\rho \pm \omega i$ and $-\rho \pm \omega i$ with $\rho \neq 0$ (see the arc denoted by $\mathscr{D}\mathscr{F}$ in Figure 5.14).

- For parameter values in between HH and HBT, the linear part has four pure imaginary eigenvalues $\pm\omega_1 i$ and $\pm\omega_2 i$ with $\omega_1 \neq \omega_2$ (see the arc denoted by $\mathscr{H}\mathscr{H}$ in Figure 5.14).

We are interested in the existence of homoclinic orbits to a bifocus. This is only possible at P_- for parameter values close to the arc $\mathscr{D}\mathscr{F}$.

As in the previous cases of nilpotent singularities, it is more convenient to consider a directional rescaling of the parameters. We use formulas as in (5.52) and in this case we assume that $v_1 = -1$ and $(v_2, v_3, v_4) = (\bar{v}_2, \bar{v}_3, \bar{v}_4) \in \mathbb{R}^3$ to obtain

$$y_2 \frac{\partial}{\partial y_1} + y_3 \frac{\partial}{\partial y_2} + y_4 \frac{\partial}{\partial y_3}$$

$$+ \left(-1 - y_2 + \bar{v}_2 y_2 + \bar{v}_3 y_3 + \bar{v}_4 y_4 + y_1^2 + \varepsilon \kappa y_1 y_2 + O(\varepsilon^2)\right) \frac{\partial}{\partial y_4}.$$

$$\tag{5.55}$$

The equilibrium points when $\varepsilon = 0$ are now $Q_\pm = (\pm 1, 0, 0, 0)$. In fact this two equilibrium points are the only equilibria for $\varepsilon > 0$ because $h(x, \mu) = O(\|(x_2, x_3, x_4)\|)$ in (5.51).

To compare with results already available in the literature we need to translate P_- to the origin. Introducing new coordinates

$$x_1 = (y_1 + 1)/2 \qquad x_2 = y_2/2^{5/4} \qquad x_3 = y_3/2^{3/2} \qquad x_4 = y_4/2^{7/4}$$

and multiplying by a factor $2^{1/4}$ we get the following family

$$
\begin{aligned}
& x_2\frac{\partial}{\partial x_1} + x_3\frac{\partial}{\partial x_2} + x_4\frac{\partial}{\partial x_3} \\
& \quad + \left(-x_1 + \eta_2 x_2 + \eta_3 x_3 + \eta_4 x_4 + x_1^2 + \bar{\varepsilon}\kappa x_1 x_2 + O(\bar{\varepsilon}^2)\right)\frac{\partial}{\partial x_4}
\end{aligned}
\tag{5.56}
$$

with

$$
\eta_2 = 2^{-3/4}(\bar{v}_2 - \varepsilon\kappa) \quad \eta_3 = 2^{-1/2}\bar{v}_3 \quad \eta_4 = 2^{-1/4}\bar{v}_4 \quad \bar{\varepsilon} = 2^{1/4}\varepsilon.
$$

The limit family restricted to the reversibility curve $\eta_2 = \eta_4 = 0$ reads as follows:

$$
x_2\frac{\partial}{\partial x_1} + x_3\frac{\partial}{\partial x_2} + x_4\frac{\partial}{\partial x_3} + \left(-x_1 + \eta_3 x_3 + x_1^2\right)\frac{\partial}{\partial x_4}.
\tag{5.57}
$$

The associated fourth order differential

$$
u^{(iv)}(t) + Pu''(t) + u(t) - u(t) = 2 = 0
\tag{5.58}
$$

where $u = x_1$ and $P = -\eta_2$ has been extensively studied in the literature because it provides the equation for the traveling waves of the one-dimensional Korteweg–de Vries equation:

$$
u_t = u_{xxxx} - bu_{xxx} + 2uu_x.
$$

In Amick and Toland (1992) authors prove that (5.58) can be written as a Hamiltonian system (as we have stated in Theorem 5.8 in a more general setting). The hypothesis required in Hofer and Toland (1984, Theorem 2) are satisfied and therefore, for each $P \leq -2$, there exists an even solution u of (5.58) with $u(t) \to 0$ when $t \to \pm\infty$ satisfying that $u(t) > 0$, $u'(t) < 0$ and $(P/2)u'(t) + u'''(t) > 0$ for all $t \in (0, \infty)$. They also prove that for all $P \leq -2$ any such even solution is unique. Note that this solution corresponds to a homoclinic orbit for system (5.57) when $\eta_3 \geq 2$. According to Buffoni, Champneys, and Toland (1996), the intersection between the invariant manifolds is transversal for the restriction to the level surface of the hamiltonian function which contains it and, consequently, it is non degenerate in the sense of Definition A.22. Moreover, again in Amick and Toland (1992), the persistence of such homoclinic solutions is argued for $P > -2$ but close enough to -2. On the other hand, in Amick and Toland (1992, Sect. 2) author checked all hypothesis required in Champneys and Toland (1993, Thm. 4.4) to conclude that a Belyakov–Devaney bifurcation takes place at $P = -2$. It consists in the emerging from the primary homoclinic solution and for each $n \in \mathbb{N}$ of a finite number of n-modal secondary homoclinics (or n-pulses) which cut n times a section transversal to the primary homoclinic orbit Belyakov (1984b), Belyakov and Shilnikov (1990), and Devaney (1976a). A more detailed description of the known catalogue of homoclinic solutions in (5.58) is provided in Barrientos, Ibáñez, and Rodríguez (2011).

Remark 5.22. *It follows that the conservative and reversible system (5.57) exhibits homoclinic orbits to the origin for parameter values close to the BD point along the curve $\mathscr{DF}$, that is, it displays bifocus homoclinic orbits (see Figure 5.14).*

We consider family (5.57) as an unfolding of the unique homoclinic orbit exhibited by the system at the Belyakov–Devaney bifurcation point

$$(\eta_2, \eta_3, \eta_4, \bar{\varepsilon}) = (0, 2, 0, 0).$$

Namely, we write (5.57) as

$$\dot{x} = f(x) + g(\lambda, x), \tag{5.59}$$

where

$$\lambda = (\lambda_1, \lambda_2, \lambda_3, \lambda_4) = (\eta_2, \eta_3 - 2, \eta_4 \varepsilon)$$

varies in a neighborhood of $(0, 0, 0, 0) \in \mathbb{R}^3$,

$$f(x_1, x_2, x_3, x_4) = (x_2, x_3, x_4, -x_1 + 2x_3 + x_1^2)$$

and

$$g(x_1, x_2, x_3, x_2) = (0, 0, 0, \lambda_1 x_2 + \lambda_2 x_3 + \lambda_3 x_4 + \lambda_4 \kappa x_1 x_2 + O(\lambda_4^2)).$$

Bifurcations exhibited when $\lambda_4 > 0$ correspond to bifurcations observed in any generic unfolding of the four-dimensional nilpotent singularity of codimension four.

We will prove that for parameter values along a surface through the origin, there exist homoclinic orbits, that is, the unfolding of the homoclinic connection exhibited for the system when $\lambda = 0$ inside family (5.57) is generic.

Note that the unperturbed system $\dot{x} = f(x)$ satisfies the following properties

(BD1) There is a first integral

$$H(x_1, x_2, x_3, x_4) = \frac{1}{2}x_1^2 - \frac{1}{3}x_1^3 - x_2^2 + x_2 x_4 - \frac{1}{2}x_3^2.$$

(BD2) The system is time reversible with respect to the involution

$$R : (x_1, x_2, x_3, x_4) \rightarrow (x_1, -x_2, x_3, -x_4).$$

(BD3) The linear part of the vector field at the equilibrium point $(0, 0, 0, 0)$ has a pair of double real eigenvalues $+1$ and -1.

(BD4) There exists a non-degenerate homoclinic orbit

$$\gamma = \{p(t) = (p_1(t), p_2(t), p_3(t), p_4(t)) : t \in \mathbb{R}\},$$

such that p_1 and p_3 are even functions, p_2 and p_4 are odd functions and, moreover, $p_1(t) > 0$, $p_2(t) < 0$ and $p_4(t) - p_2(t) > 0$ for all $t \in (0, \infty)$.

(BD5) According to Proposition A.15 and Proposition A.21, since γ is non-degenerate in the sense of Definition A.22, the dimension of the space of bounded solutions for the variational equation

$$\begin{pmatrix} \dot{z}_1 \\ \dot{z}_2 \\ \dot{z}_3 \\ \dot{z}_4 \end{pmatrix} = Df(p(t)) \begin{pmatrix} z_1 \\ z_2 \\ z_3 \\ z_4 \end{pmatrix} = \begin{pmatrix} 0 & 1 & 0 & 0 \\ 0 & 0 & 1 & 0 \\ 0 & 0 & 0 & 1 \\ -1 + 2p_1(t) & 0 & 2 & 0 \end{pmatrix} \begin{pmatrix} z_1 \\ z_2 \\ z_3 \\ z_4 \end{pmatrix}$$

and also for the adjoint variational equation

$$\begin{pmatrix} \dot{w}_1 \\ \dot{w}_2 \\ \dot{w}_3 \\ \dot{w}_4 \end{pmatrix} = -Df(p(t))^* \begin{pmatrix} w_1 \\ w_2 \\ w_3 \\ w_4 \end{pmatrix}$$

$$= \begin{pmatrix} 0 & 0 & 0 & 1 - 2p_1(t) \\ -1 & 0 & 0 & 0 \\ 0 & -1 & 0 & -2 \\ 0 & 0 & -1 & 0 \end{pmatrix} \begin{pmatrix} w_1 \\ w_2 \\ w_3 \\ w_4 \end{pmatrix}$$

equals 1. The function $f(p(t))$ is a bounded solution of the variational equation and, since the system is Hamiltonian

$$\psi(t) = \nabla H(p(t)) = (p_1(t) - p_1(t)^2, p_4(t) - 2p_2(t), -p_3(t), p_2(t)))$$

is a bounded solution of the adjoint variational equation (see Proposition 5.9).

To study the persistence of the homoclinic orbit γ we consider the splitting function $\xi^\infty : \Lambda \to \mathbb{R}$ as defined in Lemma B.1. For each λ such that $\xi^\infty(\lambda) = 0$, there is a homoclinic orbit at O.

Let

$$D_\lambda \xi^\infty(0) = \nabla \xi^\infty(0) = \left(\ \xi^\infty_{\lambda_1}(0) \quad \xi^\infty_{\lambda_2}(0) \quad \xi^\infty_{\lambda_3}(0) \quad \xi^\infty_{\lambda_4}(0) \ \right).$$

The bifurcation equation $\xi^\infty(\lambda) = 0$ can be studied by means of the Implicit Function Theorem. Namely, if $\mathrm{rank}\, D_\lambda \xi^\infty(0) = 1$, then the system has homoclinic orbits at O for parameter values on a hypersurface $\mathscr{H}$ containing $\lambda = 0$. As follows from Theorem B.2

$$\xi^\infty_{\lambda_i}(0) = \int_{-\infty}^{\infty} \left\langle \psi(t), \frac{\partial g}{\partial \lambda_i}(0, p(t)) \right\rangle dt.$$

It easily follows that

$$\xi^\infty_{\lambda_1}(0) = \int_{-\infty}^{\infty} p_2(t)^2 \, dt$$

$$\xi^\infty_{\lambda_2}(0) = \int_{-\infty}^{\infty} p_2(t) p_3(t) \, dt$$

$$\xi^\infty_{\lambda_3}(0) = \int_{-\infty}^{\infty} p_2(t) p_4(t) \, dt$$

$$\xi^\infty_{\lambda_4}(0) = \kappa \int_{-\infty}^{\infty} p_1(t) p_2(t)^2 \, dt.$$

It is straightforward that $\xi^\infty_{\lambda_1}(0) \neq 0$. Since p_2 is even and p_3 is odd, the product $p_2(t) p_3(t)$ is odd and hence $\xi^\infty_{\lambda_2}(0) = 0$. Because $p_1 > 0$ it follows that $\xi^\infty_{\lambda_4}(0) \neq 0$. Finally, taking into account that p_2 and p_4 are odd functions and integrating by parts,

$$\xi^\infty_{\lambda_3}(0) = 2 \int_{0}^{\infty} p_2(t) p_4(t) \, dt =$$

$$= \lim_{t \to \infty} p_2(t) p_3(t) - p_2(0) p_3(0) - \int_{0}^{\infty} p_3(t)^2 \, dt \neq 0.$$

Since $\xi_{\lambda_4}(0) \neq 0$, the hypersurface $\mathscr{H}$ intersects the hyperspace $\lambda_4 = 0$ transversely. The only pending question is to show that $\mathscr{H}$ contains parameter values arbitrarily close to $0 \in \Lambda$ and with $\lambda_4 > 0$ for which the system exhibits bifocal homoclinic orbits.

When working with a Belyakov–Devaney bifurcation, the equilibrium point at the origin O satisfies that $\dim W^s(O) = \dim W^u(O) = 2$ for all parameter values close enough to 0, but there four different cases must be distinguished:

- On both, $W^s(O)$ and $W^u(O)$, the equilibrium point s of focus type. We refer to this as the $\mathscr{F}\mathscr{F}$-case.

- On both, $W^s(O)$ and $W^u(O)$, the equilibrium point is of node type. We refer to this as the $\mathscr{N}\mathscr{N}$-case.

- On $W^u(O)$ and $W^s(O)$, the equilibrium point is of node and focus type, respectively. We refer to this as the $\mathscr{N}_+\mathscr{F}_-$-case.

- On $W^u(O)$ and $W^s(O)$, the equilibrium point is of focus and node type, respectively. We refer to this as the $\mathscr{F}_+\mathscr{N}_-$-case.

The characteristic polynomial of $Df(O) + D_x g(\lambda, O)$ is given by

$$Q(r, \lambda) = r^4 - a_3(\lambda)r^3 - a_2(\lambda)r^2 - a_1(\lambda)r - a_0(\lambda),$$

with

$$a_0(\lambda) = -1 + O(\lambda_4^2) \qquad a_1(\lambda) = \lambda_1 + O(\lambda_4^2)$$
$$a_2(\lambda) = 2 + \lambda_2 + O(\lambda_4^2) \qquad a_3(\lambda) = \lambda_3 + O(\lambda_4^2).$$

The condition to have a double real eigenvalue is

$$Q(r, \lambda) = D_r Q(r, \lambda) = 0.$$

Note that $(-1, 0)$ and $(1, 0)$ are solutions of the above equations. It easily follows that

$$\nabla D_r Q(1, 0) = (8, -1, -2, -3, 0)$$

and hence there exists a regular function $\hat{r}(\lambda)$ defined in a neighborhood of $\widetilde{\Lambda}$ of $\lambda = 0$ such that $r(0) = 1$ and

$$D_r Q(r(\lambda), \lambda) = 0$$

for all $\lambda \in \Lambda$. Substituting in $Q(r, \lambda) = 0$ we get the equation

$$F(\lambda) = Q(\hat{r}(\lambda), \lambda) = 0.$$

with $F(0) = 0$. Again it is straightforward that

$$\nabla F(0) = (-1, -1, -1, 0)$$

and hence, there exist a hypersurface $\mathscr{D}^+$ in the parameter space for which the equilibrium point at the origin has a double positive eigenvalue. It contains the origin of parameters where the tangent space is given by

$$\lambda_1 + \lambda_2 + \lambda_3 = 0.$$

Similarly, one can prove that there exist a hypersurface $\mathscr{D}^-$ in the parameter space for which the equilibrium point at the origin has a double negative eigenvalue. It also contains the origin of parameters an there, the tangent space is given by

$$\lambda_1 - \lambda_2 + \lambda_3 = 0.$$

Let $N = D\xi^\infty(0)$, $N_{\mathscr{D}+} = (1,1,1,0)$ and $N_{\mathscr{D}-} = (1,-1,1,0)$ be normal vectors to $\mathscr{H}$, $\mathscr{D}^+$ and $\mathscr{D}^-$ at $\lambda = 0$. Denote also $N_{\lambda_4=0} = (0,0,0,1)$ a normal vector to $\lambda_4 = 0$.

Since $\mathrm{rank}(N_{\mathscr{H}}, N_{\mathscr{D}-}, N_{\lambda_4=0}) = 3$, there exist a surface $\mathscr{H}^- = \mathscr{H} \cap D_-$ transverse to $\lambda_4 = 0$ corresponding to parameter values for which the equilibrium point at the origin in system (5.59) has a double negative eigenvalue. Also, since $\mathrm{rank}(N_{\mathscr{H}}, N_{\mathscr{D}+}, N_{\lambda_4=0}) = 3$, there exist a surface $\mathscr{H}^+ = \mathscr{H} \cap D_+$ transverse to $\lambda_4 = 0$ corresponding to parameter values for which the equilibrium point at the origin in system (5.59) has a double positive eigenvalue.

On the other hand, $\mathrm{rank}(N_{\mathscr{H}}, N_{\mathscr{D}-}, N_{\mathscr{D}+}, N_{\lambda_4=0}) = 4$ if and only if $\xi_{\lambda_1}(0) - \xi_{\lambda_3}(0) \neq 0$. Note that

$$\xi_{\lambda_1}(0) - \xi_{\lambda_3}(0) = \int_{-\infty}^{\infty} p_2(t)(p_2(t) - p_4(t))\, dt.$$

From **(BD4)** we know that p_2 and p_4 are odd functions al also that $p_2(t) < 0$ and $p_4(t) - p_2(t) < 0$ for all $t \in (0, \infty)$. Hence $p_2(p_2 - p_4)$ is an even function and $p_2(t)(p_2(t) - p_4(t)) > 0$ for all $t \in \mathbb{R}$. It follows that $\xi_{\lambda_1}(0) - \xi_{\lambda_3}(0) > 0$. Therefore, there exists a curve $\mathscr{H}^\pm = \mathscr{H} \cap \mathscr{D}^- \cap \mathscr{D}^+$ transverse to $\lambda_4 = 0$ corresponding to homoclinic orbits to a an equilibrium point with a pair of double real eigenvalues, one positive an the other negative.

Summarizing, we have proved the following result:

Theorem 5.23. *There exist a bifurcation hypersurface $\mathscr{H} \subset \mathbb{R}^4$ providing parameter values for which the system (5.59) exhibits a homoclinic orbit to the origin. On $\mathscr{H}$ there exist bifurcation surfaces $\mathscr{H}^+$ and $\mathscr{H}^-$ corresponding to parameter values for which the linear part at the origin has a double positive (resp. negative) real eigenvalue. They intersect along a bifurcation curve $\mathscr{H}^\pm$ corresponding to parameter values for which the origin has a pair of double real eigenvalues, one positive and the other negative. The union of the surfaces $\mathscr{H}^+$ and $\mathscr{H}^-$ splits $\mathscr{H}$ into four regions:*

- *$\mathscr{H}_{\mathscr{F}\mathscr{F}}$: Homoclinic orbits to a focus-focus (the bifocus case),*

- *$\mathscr{H}_{\mathscr{N}+\mathscr{F}-}$: Homoclinic orbits to a (repelling-attracting) node-focus,*

- *$\mathscr{H}_{\mathscr{F}+\mathscr{N}-}$: Homoclinic orbits to a (repelling-attracting) focus-node,*

- $\mathcal{H}_{\mathcal{N}\mathcal{N}}$: *Homoclinic orbits to a node-node.*

All these bifurcations are transverse to $\lambda_4 = 0$ and hence they are also present in any generic unfolding of the four-dimensional nilpotent singularity.

5.8 Further chaotic scenarios

In the previous two sections we have proved that chaos is generically unfolded by n-dimensional nilpotent singularities of codimension n when $n = 3$ and $n = 4$. In fact, since any n-dimensional nilpotent singularity of codimension n unfolds generically $(n - 1)$-dimensional nilpotent singularities of codimension $n - 1$ (see Drubi, Ibáñez, and Rodríguez (2007)), the result is also valid for any $n \geqslant 5$. Now we discuss the existence of lower codimension examples. We will explain how some codimension two Hopf–Zero singularities also unfold Shilnikov bifurcations and hence strange attractors. Nevertheless, when comparing this scenario with the nilpotent examples, we will point out two principal differences. First, genericity must be understood in a delicate way because it cannot be traced on any finite jet of the singularity. Second, as we will explain later, the techniques required to prove the existence of homoclinic orbits are much more involved that those used in the previous cases. Namely, a bifurcation equation is not available in the case of the unfolding of Hopf–Zero singularities.

The lowest codimension singularities in $\mathbb{R}^3$ with a three-dimensional center manifold are those ones whose linear part is linearly conjugated to

$$- \omega y \frac{\partial}{\partial x} + \omega x \frac{\partial}{\partial y}. \tag{5.60}$$

with $\omega > 0$. We refer to them as *Hopf–Zero singularities*. Classification of these singularities is due to Takens (1974b). He proved that, up to C^∞-equivalence, the 2-jet of any singularity with part lineal (5.60) can be written in the following normal form:

$$\dot{x} = -y - axz$$
$$\dot{y} = x - ayz$$
$$\dot{z} = cz^2 + b(x^2 + y^2)$$

To normalize the rotation speed ω one needs to consider a time-rescaling. It should also be noticed that this rescaling is not always considered in literature (see for instance Broer and Vegter (1984), Guckenheimer and Holmes (2002), Kuznetsov (2004)).

The degeneracy conditions imposed on the linear part and the open conditions $a \neq 0$, $b \neq 0$ and $c \neq 0$ characterize a stratum of codimension two in the space of germs of singularities of vector fields on $\mathbb{R}^3$. Coefficient c can be normalized by a scaling of coordinates and we can write (5.60) as

$$\begin{aligned}
\dot{x} &= -y - axz \\
\dot{y} &= x - ayz \\
\dot{z} &= z^2 + b(x^2 + y^2)
\end{aligned}$$

Introducing cylindrical coordinates $x = r\cos\theta$, $y = r\sin\theta$, we get $\theta' = 1$ and hence the following reduced system on the plane

$$\begin{aligned}
\dot{r} &= -arz \\
\dot{z} &= z^2 + br^2
\end{aligned} \tag{5.61}$$

Hopf-zero singularities of codimension two are 2-jet determined with respect the local topological equivalence Dumortier and Ibáñez (1998). As follows from the Takens classification Takens (1974b), there are six topological types (compare also with Guckenheimer and Holmes (2002)):

- Type I: $a > 0$ and $b > 0$,

- Type II: $a \in (-1, 0)$ and $b > 0$,

- Type III: $a < -1$ and $b > 0$,

- Type IV: $a > 0$ and $b < 0$,

- Type V: $a \in (-1, 0)$ and $b < 0$,

- Type VI: $a < -1$ and $b < 0$.

In the discussion below we are only concerned with Type I. Figure 5.15 shows the phase portrait of the reduced vector field (5.61) when $a > 0$ and $b > 0$.

Generic unfoldings of Hopf–Zero singularities can be written in the following normal form

$$\begin{aligned}
\dot{x} &= -y + \mu_2 x - axz + A(x, y, z, \mu_1, \mu_2) \\
\dot{y} &= x + \mu_2 y - ayz + B(x, y, z, \mu_1, \mu_2) \\
\dot{z} &= \mu_1 + z^2 + b(x^2 + y^2) + C(x, y, z, \mu_1, \mu_2)
\end{aligned} \tag{5.62}$$

where A, B, C are C^∞ or C^ω and of order $O(|x, y, z, \mu_1, \mu_2|^3)$. Truncating at second order and taking again cylindrical coordinates we obtain

$$\begin{aligned}
\dot{r} &= \mu_2 r - arz \\
\dot{z} &= \mu_1 + z^2 + br^2 \\
\dot{\theta} &= -1.
\end{aligned} \tag{5.63}$$

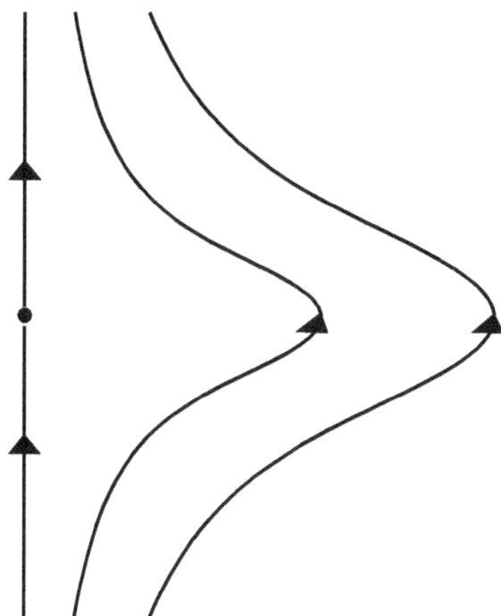

Figure 5.15: *Topological type corresponding to Hopf–Zero singularities of Type I.*

When $\mu_2 = 0$, family (5.63) has a first integral

$$H(r, z) = r^{\frac{2}{a}} \left(\mu_1 + z^2 + \frac{b}{1+a} r^2 \right). \qquad (5.64)$$

Figure 5.16 shows the phase portrait of the reduced system when $\mu_1 < 0$ and $\mu_2 = 0$ obtained from (5.63 when the θ-component of the vector field is skipped out. The truncation of (5.62) at second order has two equilibrium points $P^{\pm} = (\pm\sqrt{-\mu_1}, 0, 0)$. It follows from the study of (5.64) that $\dim W^u(P^+) = \dim W^s(P^-) = 2$. These manifolds coincide on a invariant globe contained in the set of points where $H(r, z) = 0$. Moreover the branches of the one-dimensional invariant manifolds $W^s(P^+)$ and $W^u(P^-)$ which are contained inside the globe are also coincident (see Figure 5.17).

When higher order terms are considered, all these invariant structures: the common branch shared by the one-dimensional invariant manifolds and the globe formed by the two-dimensional ones, can be destroyed. Clearly, there is a chance for the existence of Shilnikov homoclinic bifurcations, but rigorous arguments are quite involved. The first obstacle regards to the rotational symmetry exhibited by any higher-order normalization. As follows from Broer and Vegter (1984, Lemma A) such rotational symmetry implies that, under generic assumptions involving third order terms, any truncated family, written in normal form, exhibits an invariant globe for parameter values along a curve with an end at the origin. Hence, conditions for the existence of Shilnikov bifurcations

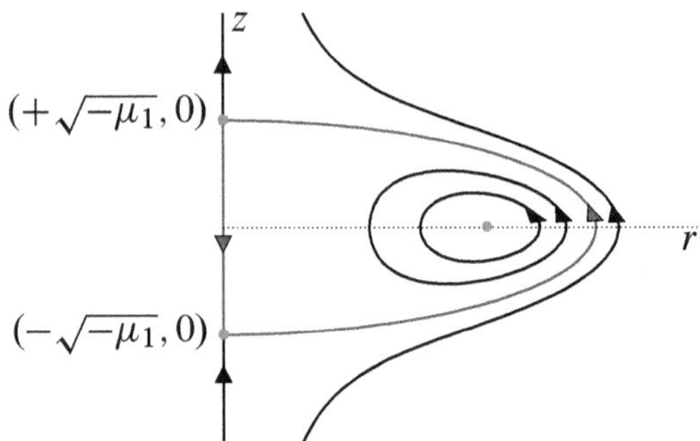

Figure 5.16: *Phase portrait of the reduced 2-jet for $\mu_1 < 0$ and $\mu_2 = 0$. There is a heteroclinic connection from $(+\sqrt{-\mu_1},0)$ to $(-\sqrt{-\mu_1},0)$ along the x-axis. This corresponds to the connection between P^+ and P^- along the one-dimensional invariant manifolds. There is another heteroclinic connection from $(-\sqrt{-\mu_1},0)$ to $(+\sqrt{-\mu_1},0)$. This corresponds to the invariant globe which contains the two-dimensional invariant manifolds through P^+ and P^-.*

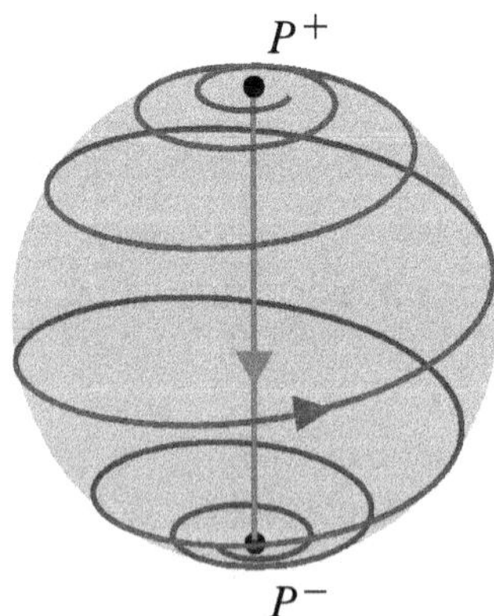

Figure 5.17: *Invariant manifolds exhibited by the family obtained by truncation of (5.62) at second order for $\mu_1 < 0$ and $\mu_2 = 0$. In red colour the connection along the one-dimensional invariant manifolds. In grey colour the invariant globe formed by the two-dimensional invariant manifolds.*

cannot be traced on any finite jet of the singularity.

The chance for the existence of Shilnikov bifurcations was already pointed out in Guckenheimer (1981) and Guckenheimer and Holmes (2002). In Gaspard (1993) a class of families of the form $X + \varepsilon Y$, where X is the second order truncation in (5.62) and ε is a small parameter, was considered. It was proved the occurrence of Shilnikov bifurcations near the codimension two point. However, it must be remarked that generic unfoldings of a Hopf–Zero singularity cannot be written in that particular form.

The question about the existence of Shilnikov bifurcations was treated in Broer and Vegter (1984) for C^{∞} unfoldings. It was proved that for any C^{∞} unfolding of a Hopf–Zero singularity there exists a flat perturbation leading to a family which exhibits the Shilnikov phenomena. Because the argument involves a flat perturbation, no usable criteria are provided and, moreover, the problem remained open for the case of analytic unfoldings.

Generic unfoldings of the Hopf–Zero singularity of Type I were considered in Dumortier, Ibánez, Kokubu, and Simó (2013). Introducing a scaling parameter $\varepsilon = \sqrt{-\mu_1}$ and scaling variables and time, one obtains either a singular perturbation problem with a pure rotation when $\varepsilon = 0$ or a family with rotation speed tending to ∞ as $\varepsilon \to 0$. In any of the two approaches there is a clear limit for the invariant manifolds of the two equilibrium points corresponding to the poles of the invariant globe. In any case, one can apply the results in Bonckaert and Fontich (2003, 2005) to prove that, when the scaling parameter tends to 0, the invariant manifolds have a limit position, given by the invariant manifolds of the equilibrium points at the 2-jet level, at least when one considers restrictions to $z \geqslant 0$ or to $z \leqslant 0$. Therefore, for any generic unfolding of the Hopf–Zero singularity of Type I, a splitting distance is well defined for both, the one-dimensional and the two-dimensional invariant manifolds. Using conjectured formulas for the splitting functions and some extra conditions, existence of Shilnikov homoclinic bifurcation points is proven for general unfoldings.

Explicit formulas for the splitting functions were obtained in Baldomá, Castejón, and Seara (2013, 2018a,b). The case of the one-dimensional splitting was solved in Baldomá, Castejón, and Seara (2013). It turns out that the distance between the one-dimensional invariant manifolds is exponentially small with respect to ε and also that the coefficient in front of the dominant term depends on the full jet of the singularity. The splitting function for the two-dimensional invariant manifolds was obtained in Baldomá, Castejón, and Seara (2018a,b). The mean free terms in the asymptotic formula are exponentially small with respect to ε and coefficients, which now they also depend on an angular variable, depend again on the full jet of the singularity. Note that because constants involved in the dominant terms depend on the full jet of the singularity, their computation can only be done by means of numerical techniques (some examples are included in Dumortier, Ibánez, Kokubu, and Simó (2013)).

Conclusive results are given in Baldomá, Ibáñez, and Seara (2019). Putting together Baldomá, Castejón, and Seara (2013, 2018a,b) and Dumortier, Ibánez, Kokubu, and Simó (2013) and with some extra work, general results for the existence of Shilnikov homoclinic bifurcations in generic analytic unfoldings of a Hopf–Zero singularity of Type I. Namely, it is proved that under generic and checkable hypothesis, any analytic unfolding of a Hopf–Zero singularity within the appropriate class contains Shilnikov homoclinic orbits, and as a consequence chaotic dynamics.

Remark 5.24. *There is yet another singularity of codimension two that is likely an organizing center of chaotic behaviors: the Hopf–Hopf singularity. Namely, we refer to four-dimensional singularities whose 1-jet is linearly conjugate to*

$$\omega_1 \left(x_1 \frac{\partial}{\partial y_1} - y_1 \frac{\partial}{\partial x_1} \right) + \omega_2 \left(x_2 \frac{\partial}{\partial y_2} - y_2 \frac{\partial}{\partial x_2} \right).$$

with $\omega_1 \neq \omega_2$. These singularities were classified in Takens (1974b) and according to Guckenheimer and Holmes (2002) and Kuznetsov (2004) there is a topological type for which, similarly to what happens with the Hopf–Zero singularity, the second order truncation of any generic unfolding exhibits a configuration which could explain the arising of chaotic dynamics. For a convenient choice of parameters, the truncated system has an equilibrium point P and two periodic orbits Γ_1 and Γ_2, such that

$$\dim W^u(P) = \dim W^s(P) = 2$$
$$\dim W^u(\Gamma_1) = \dim W^s(\Gamma_1) = 2$$
$$\dim W^u(\Gamma_2) = \dim W^s(\Gamma_2) = 2.$$

Moreover,

$$(W^u(P) \setminus \{P\}) \subset W^s(\Gamma_1),$$
$$(W^u(\Gamma_1) \setminus \{\Gamma_1\}) \subset W^s(\Gamma_2),$$
$$(W^u(\Gamma_2) \setminus \{\Gamma_2\}) \subset W^s(P).$$

Remark 5.25. *Regarding to the genesis of chaotic dynamics from local bifurcations, other singularities of great interest are the so-called Hopf–Bogdanov–Takens singularities. We are now referring to four-dimensional singularities whose 1-jet is linearly conjugated to:*

$$y_1 \frac{\partial}{\partial x_1} + \omega \left(x_2 \frac{\partial}{\partial y_2} - y_2 \frac{\partial}{\partial x_2} \right).$$

with $\omega \neq 0$. A formal classification of these singularities is described in Drubi, Ibáñez, and Rivela (2019a) at their lowest level of degeneracy: codimension three. How strange attractors are unfolded is illustrated with numerical explorations in Drubi, Ibáñez, and Rivela (2019b).

Exponential Dichotomy

A

A.1 Hyperbolic linear vector fields

A hyperbolic matrix A is a n by n real matrix whose eigenvalues all have non-zero real part. It follows that zero cannot be an eigenvalue of A, and thus every hyperbolic matrix is invertible. The transformation $f(x) = Ax$ on $\mathbb{R}^n$ is called *hyperbolic linear vector field* and the differential equation $\dot{x} = Ax$ is named *hyperbolic linear system*. The qualitative behavior of the solution of a hyperbolic linear system determines a decomposition of $\mathbb{R}^n$ into two invariant subspaces E^s and E^u usually called the *stable* and *unstable* bundle respectively Markley (2011). These subspace satisfy $\mathbb{R}^n = E^s \oplus E^u$ so that for each $x \in \mathbb{R}^n$, there is a unique decomposition $x = x_s + x_u$ with $x_s \in E^s$ and $x_u \in E^u$. Let P be the projection of $\mathbb{R}^n$ into E^s defined by $x \mapsto x_s$. Consequently, $I - P$ is the projection on E^u sending x to x_u. If $X(t)$ denotes the fundamental matrix of the system $\dot{x} = Ax$, then $X(t)X^{-1}(s)x$ with $t \in \mathbb{R}$ defines a solution so that for $t = s$ takes the value x and

$$X(t)X^{-1}(s)Px = PX(t)X^{-1}(s)x,$$
$$X(t)X^{-1}(s)(I - P)x = (I - P)X(t)X^{-1}(s)x. \tag{A.1}$$

Moreover, there are positive constants K, L, α and β such that

$$\|X(t)X^{-1}(s)P\| \leqslant Ke^{-\alpha(t-s)} \quad \text{for all } t \geqslant s,$$
$$\|X(t)X^{-1}(s)(I - P)\| \leqslant Le^{-\beta(s-t)} \quad \text{for all } s \geqslant t. \tag{A.2}$$

Both properties can be written equivalently in a common expression:

Proposition A.1. *Conditions (A.1) and (A.2) are equivalent to the existence of an idempotent matrix Q such that*

$$\|X(t)QX^{-1}(s)\| \leqslant Ke^{-\alpha(t-s)} \quad \textit{for all } t \geqslant s,$$
$$\|X(t)(I - Q)X^{-1}(s)\| \leqslant Le^{-\beta(s-t)} \quad \textit{for all } s \geqslant t. \tag{A.3}$$

Proof. Let $\mathscr{Q}(s) = X(s)QX^{-1}(s)$. Hence (A.3) is equivalent to

$$\|X(t)X^{-1}(s)\mathscr{Q}(s)\| \leqslant Ke^{-\alpha(t-s)} \quad \text{for all } t \geqslant s,$$
$$\|X(t)X^{-1}(s)(I - \mathscr{Q}(s))\| \leqslant Le^{-\beta(s-t)} \quad \text{for all } s \geqslant t. \tag{A.4}$$

In order to conclude (A.1) and (A.2) it suffices to prove that $\mathscr{Q}(\bar{s}) = P$ for all $\bar{s} \in \mathbb{R}$. But this is evident from the uniqueness of the projection on the stable and unstable subspaces: inequalities (A.4) imply that $\mathscr{Q}(\bar{s})$ and $I - \mathscr{Q}(\bar{s})$ are the projections on the stable and unstable subspace E^s and E^u respectively.

Conversely, from (A.1) follows $X^{-1}(s)PX(s) = X^{-1}(t)PX(t)$ for all $s, t \in \mathbb{R}$. Thus, it must to be a constant matrix. Denote by Q this idempotent matrix. Hence $X(t)X^{-1}(s)P = X(t)QX^{-1}(s)$ and (A.3) follows from (A.2). $\qquad\square$

The notion of hyperbolicity expressed in terms of conditions (A.3) can be extended to a non-autonomous linear differential equation of the form $\dot{x} = A(t)x$. In this context it is be called *exponential dichotomy*. We study more deeply this property in the following section. The classical references for the study of exponential dichotomies are Coppel (1978), Z. Lin and Y.-X. Lin (2000), Massera and Schäffer (1966), and Palmer (1984, 2000).

A.2   Exponential dichotomy

Let $X(t)$ be a fundamental matrix of the linear system

$$\dot{x} = A(t)x, \qquad x \in \mathbb{R}^n, \tag{A.5}$$

where $A(t)$ is defined and continuous on an interval $J \subseteq \mathbb{R}$.

Definition A.2. *System (A.5) has an exponential dichotomy on J if there is a projection $P : \mathbb{R}^n \to \mathbb{R}^n$ (an n by n matrix P with $P^2 = P$) and positive constants K, L, α and β such that for every $s, t \in J$,*

$$\|X(t)PX^{-1}(s)\| \leqslant Ke^{-\alpha(t-s)} \quad \textit{for } t \geqslant s,$$
$$\|X(t)(I - P)X^{-1}(s)\| \leqslant Le^{-\beta(s-t)} \quad \textit{for } s \geqslant t. \tag{A.6}$$

This definition is independent of the fundamental matrix $X(t)$. Indeed, for any other fundamental matrix $Y(t)$ there is a non-singular matrix M of constant coefficients such that $Y(t) = X(t)M$. In this way, taking $Q = M^{-1}PM$ we have a new projection which allows us to replace $P = MQM^{-1}$ in (A.6) to obtain

$$\|Y(t)QY^{-1}(s)\| \leqslant Ke^{-\alpha(t-s)} \quad \text{for } t \geqslant s,$$

$$\|Y(t)(I - Q)Y^{-1}(s)\| \leqslant Le^{-\beta(s-t)} \quad \text{for } s \geqslant t.$$

Let us define $\mathscr{P}(s) = X(s)PX^{-1}(s)$ for each $s \in J$. Notice that, according with the above definition, $\mathscr{P}(s)$ is the projection corresponding to the fundamental matrix $Y(t) = X(t)X^{-1}(s)$ of (A.5) and we can give an alternative definition of exponential dichotomy.

Definition A.3. *System* (A.5) *has an exponential dichotomy on J if for all $s \in J$ there is a projection $\mathscr{P}(s) : \mathbb{R}^n \to \mathbb{R}^n$ and positive constants K, L, α and β independents of s such that for all $t \in J$ the matrix $X^{-1}(t)\mathscr{P}(t)X(t)$ has constant coefficients and*

$$\|X(t)X^{-1}(s)\mathscr{P}(s)\| \leqslant Ke^{-\alpha(t-s)} \quad \textit{for all } t \geqslant s,$$

$$\|X(t)X^{-1}(s)(I - \mathscr{P}(s))\| \leqslant Le^{-\beta(s-t)} \quad \textit{for all } s \geqslant t.$$

The continuity of the projection follows from its definition as $\mathscr{P}(s) = X(s)PX^{-1}(s)$. On the other hand, taking $s = t$ in Definition A.2, $\mathscr{P}(t)$ and $I - \mathscr{P}(t)$ are both uniformly bonded for all $t \in J$. This actually means that the angle between the subspace $\mathcal{R}(\mathscr{P}(t))$ and $\mathcal{N}(\mathscr{P}(t))$, range and kernel of $\mathscr{P}(t)$ respectively, remains uniformly bounded away from zero for all $t \in J$. In order to justify this observation we introduce the notion of angle between complementary subspaces as follows:

Definition A.4. *Let E and F be vector subspaces of $\mathbb{R}^n$ such that $\mathbb{R}^n = E \oplus F$. Consider the unique linear map $L : E^\perp \to E$ so that $F = \{v + Lv : v \in E^\perp\}$ where $E^\perp$ denotes the orthogonal space to E. Finally, the angle between E and F is defined as* $\text{ang}(E, F) = \|L\|^{-1}$.

Notice that the above definition of angle is an approximation of the geometrical notion of smaller angle between subspaces. Indeed, assume that E and F are two one-dimensional subspaces in $\mathbb{R}^2$ with a small geometric angle α and take $v \in E^\perp$ with $|v| = 1$. Then

$$\alpha \approx \tan \alpha = \frac{|v|}{|Lv|} \geqslant \left(\sup\{ \frac{|Lv|}{|v|} : |v| = 1 \} \right)^{-1} = \|L\|^{-1}.$$

The following proposition follows by observing that $\mathscr{P}(t)$ projects on $\mathcal{R}(\mathscr{P}(t))$ along the direction $\mathcal{N}(\mathscr{P}(t))$ as Figure A.1 shows:

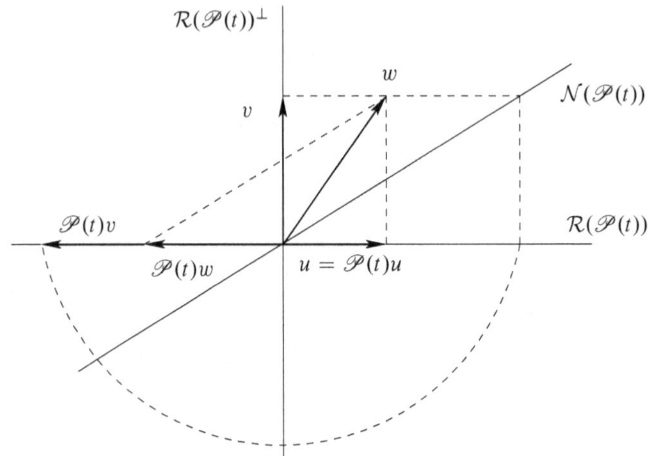

Figure A.1: *Geometric interpretation of the projection $\mathscr{P}(t)$.*

Proposition A.5. *If the linear system (A.5) has an exponential on J with projection $\mathscr{P}(t)$ and constant $K > 0$ as in Definition A.3 then*

$$\operatorname{ang}(\mathcal{R}(\mathscr{P}(t)), \mathcal{N}(\mathscr{P}(t))) \geq K^{-1} \quad \textit{for all } t \in J.$$

Proof. It suffices to note that

$$\mathcal{N}(\mathscr{P}(t)) = \mathcal{R}(I - \mathscr{P}(t)) = \{v - \mathscr{P}(t)v : v \in \mathcal{R}(\mathscr{P}(t))^{\perp}\}.$$

Thus, $\operatorname{ang}(\mathcal{R}(\mathscr{P}(t)), \mathcal{N}(\mathscr{P}(t))) = \| - \mathscr{P}(t)\|^{-1} \geq K^{-1}$.   □

Although the notion of exponential dichotomy is stated for any $J \subseteq \mathbb{R}$, the most interesting cases are when J is not bounded. We are particularly interested in $J = [\tau, \infty)$ or $J = (-\infty, \tau]$. In such cases the notions of stable and unstable subspaces can be introduced in terms of the ranges of the projections of the exponential dichotomies.

Definition A.6. *Suppose $J = [\tau, \infty)$ (resp. $J = (-\infty, \tau]$) in system (A.5). For each $t_0 \in J$ the stable (resp. unstable) subspace for initial time $t = t_0$ is defined as*

$$E_{t_0}^s = \{\xi \in \mathbb{R}^n : |X(t)X^{-1}(t_0)\xi| \to 0 \text{ as } t \to \infty\}$$
$$(\textit{resp. } E_{t_0}^u = \{\xi \in \mathbb{R}^n : |X(t)X^{-1}(t_0)\xi| \to 0 \text{ as } t \to -\infty\}).$$

In the following proposition we relate the stable and unstable subspaces with the range and kernel of the projection in the exponential dichotomy.

Proposition A.7. *Assuming that system* (A.5) *has an exponential dichotomy on J then*

i) *if $J = [\tau, \infty)$, $E_{t_0}^s$ coincides with the range $\mathcal{R}(\mathscr{P}(t_0))$ of $\mathscr{P}(t_0)$ for all $t_0 \in J$. Furthermore*

$$\mathcal{R}(\mathscr{P}(t_0)) = \{\xi \in \mathbb{R}^n : \sup_{t \geq t_0} |X(t)X^{-1}(t_0)\xi| < \infty\},$$

and for all $t_0, t_1 \in J$ it follows that $E_{t_1}^s = X(t_1)X^{-1}(t_0)E_{t_0}^s$.

ii) *if $J = (-\infty, \tau]$, $E_{t_0}^u$ coincides with the kernel $\mathcal{N}(\mathscr{P}(t_0))$ of $\mathscr{P}(t_0)$ for all $t_0 \in J$. Furthermore*

$$\mathcal{N}(\mathscr{P}(t_0)) = \{\xi \in \mathbb{R}^n : \sup_{t \leq t_0} |X(t)X^{-1}(t_0)\xi| < \infty\},$$

and for all $t_0, t_1 \in J$ it follows that $E_{t_1}^u = X(t_1)X^{-1}(t_0)E_{t_0}^u$.

Proof. According to Definitions A.6 and A.3 it follows that

$$\mathcal{R}(\mathscr{P}(t_0)) \subseteq E_{t_0}^s \subseteq \{\xi \in \mathbb{R}^n : \sup_{t \geq t_0} |X(t)X^{-1}(t_0)\xi| < \infty\}.$$

To prove the reciprocal inclusion, take $\xi \in \mathbb{R}^n$ so that

$$\sup_{t \geq t_0} |X(t)X^{-1}(t_0)\xi| < \infty.$$

Then for every $t \geq t_0$,

$$
\begin{aligned}
|(I - \mathscr{P}(t_0))\xi| &= |(I - \mathscr{P}(t_0))X(t_0)X^{-1}(t)X(t)X^{-1}(t_0)\xi| \\
&= |X(t_0)X^{-1}(t)(I - \mathscr{P}(t))X(t)X^{-1}(t_0)\xi| \\
&\leq Le^{-\beta(t-t_0)}|X(t)X^{-1}(t_0)\xi|,
\end{aligned}
$$

which tends to zero as $t \to \infty$. Hence $(I - \mathscr{P}(t_0))\xi = 0$ and thus we have $\xi \in \mathcal{R}(\mathscr{P}(t_0))$.

On the other hand, since $x \in E_{t_0}^s$ if and only if $\mathscr{P}(t_0)x = x$,

$$\mathscr{P}(t_1)X(t_1)X^{-1}(t_0)x = X(t_1)X^{-1}(t_0)\mathscr{P}(t_0)x = X(t_1)X^{-1}(t_0)x.$$

This is equivalent to $X(t_1)X^{-1}(t_0)x \in E_{t_1}^s$ and, consequently it holds $E_{t_1}^s = X(t_1)X^{-1}(t_0)E_{t_0}^s$. By similar arguments one can show the second item and thus we conclude the proof. $\square$

From the above proposition it follows that the linear flow sends $E_{t_0}^s$ and $E_{t_0}^u$ to $E_{t_1}^s$ and $E_{t_1}^u$, respectively. Accordingly, once $E_{t_0}^s$ and $E_{t_0}^u$ are fixed, the stable and unstable subspaces are determined for all t. Therefore, the projections are also determined for each $t \in J$ once they are defined for $t = t_0$. The same observation follows taking into account the uniqueness of solutions for the equation

$$\mathscr{P}'(s) = X'(s)PX^{-1}(s) + X(s)P(X^{-1}(s))' = A(s)\mathscr{P}(s) - \mathscr{P}(s)A(s).$$

Remark A.8. *The existence of an exponential dichotomy on $(-\infty, \infty)$ determines univocally P and $I - P$ as the respective projections on the stable and unstable subspaces. However, the projections for a dichotomy on $[\tau, \infty)$ or on $(-\infty, \tau]$ are not univocally determined. This is because one of the conditions in (A.6) is verified independently of the choice of $I - P$ or P.*

The following result collects the difference between an exponential dichotomy on $[\tau, \infty)$ or $(-\infty, \tau]$ and on $(-\infty, \infty)$.

Lemma A.9. *Assume system (A.5) with $J = (-\infty, \infty)$.*

   i) If (A.5) has an exponential dichotomy on the interval $[\tau, \infty)$ (resp. on $(-\infty, \tau]$) for some $\tau \in \mathbb{R}$ then it has exponential dichotomy on $[t_0, \infty)$ (resp. on $(-\infty, t_0]$) for all $t_0 \in \mathbb{R}$.

   ii) The system (A.5) has exponential dichotomy on $(-\infty, \infty)$ if and only if it has an exponential dichotomy on both intervals, $[t_0, \infty)$ and $(-\infty, t_0]$, and its corresponding stable and unstable subspaces satisfy $\mathbb{R}^n = E_{t_0}^s \oplus E_{t_0}^u$.

Proof. Suppose that (A.5) has an exponential dichotomy on $[\tau, \infty)$. The proof for $(-\infty, \tau]$ is analogous. It is clear that for every $t_0 \geq \tau$ system (A.5) has an exponential dichotomy on $[t_0, \infty)$. If $t_0 < \tau$, by Grönwall's lemma follows

$$\|X(t)X^{-1}(s)\| \leq e^{\int_s^t \|A(u)\|\, du} \quad \text{for any } s \leq t.$$

In particular, for $t_0 \leq t, s \leq \tau$ it holds

$$\|X(t)X^{-1}(s)\| \leq e^{\int_{t_0}^{\tau} \|A(u)\|\, du} = N(t_0).$$

If $t_0 \leq s \leq \tau \leq t$ then

$$\|X(t)PX^{-1}(s)\| \leq \|X(t)PX^{-1}(\tau)\|N(t_0)$$
$$\leq N(t_0)Ke^{-\alpha(t-\tau)}$$
$$\leq N(t_0)Ke^{-\alpha(t_0-\tau)}e^{-\alpha(t-s)}.$$

If $t_0 \leqslant s \leqslant t \leqslant \tau$ then

$$\|X(t)PX^{-1}(s)\| \leqslant N(t_0)^2 \|X(\tau)PX^{-1}(\tau)\| \leqslant N(t_0)^2 K.$$

Hence, in both inequalities we conclude

$$\|X(t)PX^{-1}(s)\| \leqslant \widetilde{K}(t_0)e^{-\alpha(t-s)}$$

for any $t, s \in [t_0, \infty)$ with $t \geqslant s$ where $\widetilde{K}(t_0) = N(t_0)^2 K e^{-\alpha(t_0-\tau)}$. In the same way, we also get

$$\|X(t)(I - P)X^{-1}(s)\| \leqslant \widetilde{L}(t_0)e^{-\beta(s-t)}$$

for any $t, s \in [t_0, \infty)$ with $s \geqslant t$ where $\widetilde{L}(t_0) = N(t_0)^2 L e^{-\beta(t_0-\tau)}$.

On the other hand, if the system (A.5) has exponential dichotomy on $(-\infty, \infty)$ then for every $t_0 \in \mathbb{R}$ it has an exponential dichotomy on both intervals, $[t_0, \infty)$ and $(-\infty, t_0]$, with the same projection $\mathscr{P}(t_0)$. Thus, $\mathbb{R}^n = \mathcal{R}(\mathscr{P}(t_0)) \oplus \mathcal{N}(\mathscr{P}(t_0)) = E_{t_0}^s \oplus E_{t_0}^u$.

Reciprocally, suppose that (A.5) has an exponential dichotomy on both, $[t_0, \infty)$ and $(-\infty, t_0]$. Moreover, assume that the corresponding stable and unstable subspaces satisfy that $\mathbb{R}^n = E_{t_0}^s \oplus E_{t_0}^u$. Then, there are projections $P : \mathbb{R}^n \to E_{t_0}^s$ and $I - Q : \mathbb{R}^n \to E_{t_0}^u$ such that

$$\|X(t)PX^{-1}(s)\| \leqslant Ke^{-\alpha(t-s)} \quad \text{for } t \geqslant s,$$

$$\|X(t)(I - P)X^{-1}(s)\| \leqslant Le^{-\beta(s-t)} \quad \text{for } s \geqslant t,$$

with $t, s \geqslant t_0$ and

$$\|X(t)QX^{-1}(s)\| \leqslant \widetilde{K}e^{-\widetilde{\alpha}(t-s)} \quad \text{for } t \geqslant s,$$

$$\|X(t)(I - Q)X^{-1}(s)\| \leqslant \widetilde{L}e^{-\widetilde{\beta}(s-t)} \quad \text{for } s \geqslant t.$$

with $t, s \leqslant t_0$. If $P = Q$ we get exponential dichotomy on $(-\infty, \infty)$. However, in general, it might not be like that. Nevertheless, since $\mathbb{R}^n = E_{t_0}^s \oplus E_{t_0}^u$, we will prove that there is a projection $\hat{P}$ such that

$$\|X(t)\hat{P}X^{-1}(s)\| \leqslant \hat{K}e^{-\alpha(t-s)} \quad \text{for } t \geqslant s,$$

$$\|X(t)(I - \hat{P})X^{-1}(s)\| \leqslant \hat{L}e^{-\ddot{\beta}(s-t)} \quad \text{for } s \geqslant t, \tag{A.7}$$

with $t, s \in (-\infty, \infty)$. Indeed, let $\hat{P} : \mathbb{R}^n \to E_{t_0}^s$ be the projection whose kernel is $E_{t_0}^u$. Since $\mathcal{R}(\hat{P}) = \mathcal{R}(P)$ then $\hat{P}P = P$ and $P\hat{P} = \hat{P}$. Hence,

$$P - \hat{P} = P(P - \hat{P}) = (P - \hat{P})(I - P).$$

Thus, for $s, t \geqslant t_0$

$$\|X(t)(P - \hat{P})X^{-1}(s)\| = \|X(t)PX^{-1}(t_0)X(t_0)(P - \hat{P})X^{-1}(s)\|$$
$$\leqslant Ke^{-\alpha(t-t_0)}\|X(t_0)(P - \hat{P})X^{-1}(s)\|$$
$$\leqslant Ke^{-\alpha(t-t_0)}\|X(t_0)(P - \hat{P})X^{-1}(t_0)\|\|X(t_0)(I - P)X^{-1}(s)\|$$
$$\leqslant KLe^{-\alpha(t-t_0)}e^{-\beta(s-t_0)}\|X(t_0)(P - \hat{P})X^{-1}(t_0)\|.$$

It follows that for $t \geqslant s \geqslant t_0$,

$$\|X(t)\hat{P}X^{-1}(s)\| \leqslant \|X(t)PX^{-1}(s)\| + \|X(t)(P - \hat{P})X^{-1}(s)\|$$
$$\leqslant (1 + L\|X(t_0)(P - \hat{P})X^{-1}(t_0)\|)Ke^{-\alpha(t-s)}.$$

Similarly, for $s \geqslant t \geqslant t_0$

$$\|X(t)(I - \hat{P})X^{-1}(s)\| \leqslant (1 + K\|X(t_0)(P - \hat{P})X^{-1}(t_0)\|)Le^{-\beta(s-t)}.$$

Therefore, (A.5) has an exponential dichotomy on $[t_0, \infty)$ with projection $\hat{P}$, the exponents α, β being unaltered and new constants K and L multiplied by a factor $1 + L\|X(t_0)(P - \hat{P})X^{-1}(t_0)\|$ and $1 + K\|X(t_0)(P - \hat{P})X^{-1}(t_0)\|$ respectively. On the other hand, since $\mathcal{N}(\hat{P}) = \mathcal{N}(Q)$ by a similar argument we get that (A.5) has also an exponential dichotomy on $(-\infty, t_0]$ with the same projection $\hat{P}$, unaltered exponents $\tilde{\alpha}$, $\tilde{\beta}$ and similar constants multiplied by a factor. Therefore we get (A.7) and so the dichotomy on $(-\infty, \infty)$. $\qquad\square$

Although in general, as stated in the previous proposition, having an exponential dichotomy on an interval of semi-infinite length is not enough to have an exponential dichotomy on the entire real line. However, in the particular case that the matrix $A(t)$ is periodic these notions are equivalent:

Proposition A.10. *Assume that the linear system* (A.5) *is also periodic. Then, the following statements are equivalent:*

   i) System (A.5) *has exponential dichotomy on* $(-\infty, \infty)$,

   ii) System (A.5) *has an exponential dichotomy on* $[t_0, \infty)$,

   iii) System (A.5) *has an exponential dichotomy on* $(-\infty, t_0]$,

   iv) Floquet multiplies of (A.5) *have different module from one.*

Proof. The equivalence between (i) and (iv) follows from Floquet's theory Chicone (2006): there is a time-dependent change of coordinates, which transforms the periodic system $\dot{x} = A(t)x$ in a linear system of constant coefficients.

Therefore, we have an exponential dichotomy on $(-\infty, \infty)$ if and only if this autonomous linear system is hyperbolic. In turn, this is equivalent to having all the Floquet multipliers lie off the unit circle. On the other hand, (i) implies (ii) and (iii).

Reciprocally, suppose that (A.5) has an exponential dichotomy on $[t_0, \infty)$ with projection P. The case $(-\infty, t_0]$ is followed in a similar way. The translated equation $\dot{x} = A(t + kT)x$ has for fundamental matrix $X_k(t) = X(t + kT)X^{-1}(kT)$ where $T > 0$ and $k \in \mathbb{N}$ large enough. Denoting $P_k = \mathscr{P}(kT) = X(kT)PX^{-1}(kT)$, it follows that

$$\|X_k(t)P_k X_k^{-1}(s)\| \le K e^{-\alpha(t-s)} \quad \text{for } t \ge s \ge t_0 - kT,$$

$$\|X_k(t)(I - P_k)X_k^{-1}(s)\| \le L e^{-\beta(s-t)} \quad \text{for } s \ge t \ge t_0 - kT.$$

Hence $\|P_k\| \le K$ and thus, there is a subsequence $P_{k_j} \to Q$, where Q is also a projection. By taking T as the period of $A(t)$, we actually have that $X_k(t) = X(t)X^{-1}(0)$. Thus, $X_k(t) \to X(t)X^{-1}(0) = Y(t)$ and for any $t, s \in (-\infty, \infty)$ it holds

$$\|Y(t)QY^{-1}(s)\| \le K e^{-\alpha(t-s)} \quad \text{for } t \ge s,$$

$$\|Y(t)(I - Q)Y^{-1}(s)\| \le L e^{-\beta(s-t)} \quad \text{for } s \ge t,$$

Since $Y(t)$ is a fundamental matrix of (A.5) then the above inequalities imply that (A.5) has exponential dichotomy on $(-\infty, \infty)$. $\qquad\square$

The next result Palmer (2000, Lemma 7.4) states that exponential dichotomy is a robust property by perturbing small enough $A(t)$.

Proposition A.11. *Suppose that (A.5) has an exponential dichotomy on J with projection matrix function $\mathscr{P}(t)$, constants K_1, K_2 and exponents α_1, α_2. Let β_1 and β_2 be such that $0 < \beta_1 < \alpha_1$ and $0 < \beta_2 < \alpha_2$. Then there exists $\delta_0 = \delta_0(K_1, K_2, \alpha_1, \alpha_2, \beta_1, \beta_2) > 0$ such that if $B(t)$ is a continuous matrix function with*

$$\|B(t)\| \le \delta_t \le \delta_0 \quad \text{for all } t \in J,$$

the perturbed system

$$x' = [A(t) + B(t)]x$$

has an exponential dichotomy on J with constants L_1, L_2 exponents β_1, β_2 and projection matrix $\mathscr{Q}(t)$ satisfying that

$$\|\mathscr{Q}(t) - \mathscr{P}(t)\| \le N\delta_t,$$

where L_1, L_2, N are constants depending only on K_1, K_2, α_1, α_2.

As we advanced in Remark A.8, the projections of an exponential dichotomy on a half-bounded interval are not univocally determined. That is, there is no uniqueness when choosing the complement in $\mathbb{R}^n$ to the range or the kernel of the projection. One can choose the orthogonal complement. The study of the exponential dichotomy for the adjoint equation, which will be done in the next section, will allow to establish this orthogonal complement.

A.3　Dichotomy for the adjoint equation

Let $X(t)$ be a fundamental matrix of the linear equation (A.5)

$$\dot{x} = A(t)x, \qquad x \in \mathbb{R}^n, \ t \in J.$$

It is easy to see that the transpose of its inverse $X^{-1}(t)^*$ is a fundamental matrix of the *adjoint equation*

$$\dot{w} = -A(t)^* w, \qquad x \in \mathbb{R}^n, \ t \in J \tag{A.8}$$

where $A(t)^*$ denotes the transposed matrix of $A(t)$. From this relation between the fundamental matrices of both equations we can conclude the following result:

Proposition A.12. *If the equation* (A.5) *has an exponential dichotomy on J with projection matrix $\mathscr{P}(t)$ then the adjoint equation* (A.8) *has exponential dichotomy on J with projection matrix $I - \mathscr{P}(t)^*$. Moreover, for each $t_0 \in J$*

$$\mathbb{R}^n = \mathcal{R}(\mathscr{P}(t_0)) \perp \mathcal{R}(I - \mathscr{P}(t_0)^*) = \mathcal{R}(\mathscr{P}(t_0)) \perp \mathcal{N}(\mathscr{P}(t_0)^*),$$
$$\mathbb{R}^n = \mathcal{R}(I - \mathscr{P}(t_0)) \perp \mathcal{R}(\mathscr{P}(t_0)^*) = \mathcal{N}(\mathscr{P}(t_0)) \perp \mathcal{R}(\mathscr{P}(t_0)^*).$$

Proof. Taking transposed matrix in Definition A.2,

$$\|X^{-1}(t)^*(I - P^*)X(s)^*\| \leqslant Le^{-\beta(t-s)} \quad \text{para } t \geqslant s,$$
$$\|X^{-1}(t)^* P^* X(s)^*\| \leqslant Ke^{-\alpha(s-t)} \quad \text{para } s \geqslant t,$$

with $s, t \in J$. Since $X^{-1}(t)^*$ is a fundamental matrix of (A.8), we get that this equation has and exponential dichotomy on J with projection $Q = I - P^*$. On the other hand,

$$< \mathscr{P}(t_0)x, (I - \mathscr{P}(t_0)^*)w >= x^* \mathscr{P}(t_0)^*(I - \mathscr{P}(t_0)^*)w = 0.$$

Thus, $\mathcal{R}(I - \mathscr{P}(t_0)^*) \subseteq \mathcal{R}(\mathscr{P}(t_0))^{\perp}$. In order to prove the equality and conclude then that $\mathbb{R}^n = \mathcal{R}(\mathscr{P}(t_0)) \perp \mathcal{R}(I - \mathscr{P}(t_0)^*)$, it suffices taking into account that $\dim \mathcal{R}(I - \mathscr{P}(t_0)^*) = \dim \mathcal{R}(I - \mathscr{P}(t_0))$. Therefore,

$$\dim \mathcal{R}(\mathscr{P}(t_0)) + \dim \mathcal{R}(I - \mathscr{P}(t_0)^*)$$
$$= \dim \mathcal{R}(\mathscr{P}(t_0)) + \dim \mathcal{N}(\mathscr{P}(t_0)) = n.$$

Same arguments conclude that $\mathbb{R}^n = \mathcal{R}(I - \mathscr{P}(t_0)) \perp \mathcal{R}(\mathscr{P}(t_0)^*)$. $\qquad\square$

As done in Definition A.6 we can define now the stable and unstable subspaces for adjoint equations.

Definition A.13. *Suppose that $J = [\tau, \infty)$ (resp. $J = (-\infty, \tau])$ in system (A.5). For each $t_0 \in J$ the stable (resp. unstable) subspace for initial time $t = t_0$ of the adjoint equation (A.8) is defined as*

$$E_{t_0}^{s*} = \{w \in \mathbb{R}^n : |X^{-1}(t)^* X(t_0)^* w| \to 0 \text{ when } t \to \infty\}$$

$$(resp. \ E_{t_0}^{u*} = \{w \in \mathbb{R}^n : |X^{-1}(t)^* X(t_0)^* w| \to 0 \text{ when } t \to -\infty\}).$$

The following result about the relationship between the invariant subspaces of the equation $\dot{x} = A(t)x$ and its adjoint follows as a straight consequence of Proposition A.7 and Proposition A.12.

Proposition A.14. *Suppose that the linear system (A.5) has an exponential dichotomy in J.*

1. *If $J = [t_0, \infty)$ then*

$$E_{t_0}^{s} = \mathcal{R}(\mathscr{P}(t_0)) = \{x \in \mathbb{R}^n : \sup_{t \geq t_0} |X(t)X^{-1}(t_0)x| < \infty\},$$

$$E_{t_0}^{s*} = \mathcal{N}(\mathscr{P}(t_0)^*) = \{w \in \mathbb{R}^n : \sup_{t \geq t_0} |X^{-1}(t)^* X(t_0)^* w| < \infty\},$$

and $\mathbb{R}^n = E_{t_0}^{s} \perp E_{t_0}^{s*}$.

2. *If $J = (-\infty, t_0]$ then*

$$E_{t_0}^{u} = \mathcal{N}(\mathscr{P}(t_0)) = \{x \in \mathbb{R}^n : \sup_{t \leq t_0} |X(t)X^{-1}(t_0)x| < \infty\},$$

$$E_{t_0}^{u*} = \mathcal{R}(\mathscr{P}(t_0)^*) = \{w \in \mathbb{R}^n : \sup_{t \leq t_0} |X^{-1}(t)^* X(t_0)^* w| < \infty\},$$

and $\mathbb{R}^n = E_{t_0}^{u} \perp E_{t_0}^{u*}$.

In short, if the linear equation $\dot{x} = A(t)x$ has exponential dichotomy in $J = [t_0, \infty)$ (resp. $(-\infty, t_0]$) then the forward (resp. backward) bounded solutions of this equation and its adjoint are those which tend to zero exponentially when $t \to \infty$ (resp. $t \to -\infty$). On the other hand, from the decompositions of $\mathbb{R}^n$ given in Proposition A.14 it follows that, if $\dot{x} = A(t)x$ has m linearly independent forward (resp. backward) bounded solutions, then the adjoint equation $\dot{w} = -A(t)^* w$ has $n-m$ linearly independent forward (resp. backward) bounded solutions. Hence, by denoting

$$E_{t_0}^{*} = E_{t_0}^{s*} \cap E_{t_0}^{u*} = [E_{t_0}^{s} + E_{t_0}^{u}]^{\perp}$$

we obtain the following result:

Proposition A.15. *If the linear equation* (A.5) *has an exponential dichotomy in* $[t_0, \infty)$ *and in* $(-\infty, t_0]$ *then the number of linearly independent bounded solutions of the adjoint equation* (A.8) *is*

$$\dim E_{t_0}^* = n - \dim E_{t_0}^s - \dim E_{t_0}^u + \dim E_{t_0}^s \cap E_{t_0}^u.$$

In the following section, we will use knowledge on exponential dichotomy of the linear system (A.5) to characterize the of bounded solutions of the complete linear equation. This characterization will be useful in the next chapter where we study the bifurcation equation of (homo)heteroclinic connection of a non-linear differential equation.

A.4　Complete linear equation

Consider the complete linear equation

$$\dot{x} = A(t)x + b(t), \qquad x \in \mathbb{R}^n, \ t \in J \tag{A.9}$$

where $b(t)$ belongs to the Banach space $C_b^r(J, \mathbb{R}^n)$ of bounded continuous $\mathbb{R}^n$-valued functions whose derivatives up to order r exist and are bounded and continuous. We are interesting to study the bounded solutions of (A.9). To this goal, the following theorem proved in Palmer (1984, lem. 4.2) and Palmer (1988) establishes the existence of bounded solutions for the complete linear equation from the existence of bounded solutions for the adjoint equation (A.9). First, recall that a linear operator is said to be *Fredholm* if its kernel is finite dimensional and its range is closed with finite codimension. The difference between the dimension of the kernel and the codimension of the range is called *index*.

Theorem A.16. *Let* $A(t)$ *be a bounded and continuous matrix on* $(-\infty, \infty)$. *The linear equation* (A.5) *has an exponential dichotomy on* $[t_0, \infty)$ *and on* $(-\infty, t_0]$ *if and only if the linear operator*

$$L : x(t) \in C_b^1(\mathbb{R}, \mathbb{R}^n) \mapsto \dot{x}(t) - A(t)x(t) \in C_b^0(\mathbb{R}, \mathbb{R}^n)$$

is Fredholm. The index of L *is* $\dim E_{t_0}^s + \dim E_{t_0}^u - n$. *Moreover,* $b \in \mathcal{R}(L)$ *if and only if*

$$\int_{-\infty}^{\infty} <w(t), b(t)> \, dt = 0$$

for all bounded solutions $w(t)$ *of the adjoint equation* (A.8).

Assume, for instance, that the linear equation $\dot{x} = A(t)x$ has exponential dichotomy on $(-\infty, \infty)$ being $A(t)$ a bounded and continuous matrix defined on the real line. Then $\mathbb{R}^n = E_{t_0}^s \oplus E_{t_0}^u$ and from this follows that the index of L in Theorem A.16 is zero. Moreover, the constant solution $w = 0$ is the unique bounded solution of the adjoint equation $\dot{w} = -A(t)^* w$. Therefore, $\mathcal{R}(L) = C_b^0(\mathbb{R}, \mathbb{R}^n)$, and hence, the kernel of L is the trivial space zero. Consequently, for every $b(t)$ in $C_b^0(\mathbb{R}, \mathbb{R}^n)$ there is a unique bounded solution $x(t)$ of the complete linear equation $\dot{x} = A(t)x + b(t)$.

However, the above situation is not the case if the linear equation $\dot{x} = A(t)x$ has only an exponential dichotomy on $[t_0, \infty)$. The following lemma characterize the positive bounded solutions of the complete linear equation in this case.

Lemma A.17. *Assume that the linear equation* (A.5) *has an exponential dichotomy on* $J = [t_0, \infty)$. *Let* $b \in C_b^0(J, \mathbb{R}^n)$. *Then,* $x^+(t)$ *is a positive bounded solution of* (A.9) *if and only if*

$$x^+(t) = X(t)X^{-1}(t_0)\mathcal{P}(t_0)x^+(t_0)$$

$$+ \int_{t_0}^{t} X(t)X^{-1}(s)\mathcal{P}(s)b(s)\,ds \tag{A.10}$$

$$- \int_{t}^{\infty} X(t)X^{-1}(s)(I - \mathcal{P}(s))b(s)\,ds.$$

Proof. Since $X(t)$ is the fundamental matrix of (A.5), the solutions of the complete linear equation (A.9) are of the form

$$x(t) = X(t)X^{-1}(t_0)x(t_0) + X(t)\int_{t_0}^{t} X^{-1}(s)b(s)\,ds.$$

By means of the projection P associated with the exponential dichotomy of (A.5) on $J = [t_0, \infty)$, these solutions can be written as

$$x(t) = X(t)PX^{-1}(t_0)x(t_0) + X(t)(I - P)X^{-1}(t_0)x(t_0) \tag{A.11}$$

$$+ X(t)\int_{t_0}^{t} PX^{-1}(s)b(s)\,ds + X(t)\int_{t_0}^{t} (I - P)X^{-1}(s)b(s)\,ds.$$

On the other hand, according to the exponential dichotomy

$$\|X(t)PX^{-1}(s)\| \leqslant Ke^{-\alpha(t-s)} \qquad \text{for } t \geqslant s \geqslant t_0,$$

it follows that

$$|X(t)PX^{-1}(t_0)x(t_0)| \leqslant Ke^{-\alpha(t-t_0)}|x(t_0)|$$

$$\left|X(t)\int_{t_0}^{t} PX^{-1}(s)b(s)\,ds\right| \leqslant \int_{t_0}^{t} Ke^{-\alpha(t-s)}|b(s)|\,ds.$$

Thus, the first and third term of (A.11) are bounded for $t \geq t_0$.

If we assume that $x(t)$ is bounded, necessarily then the sum

$$X(t)(I - P)X^{-1}(t_0)x(t_0) + X(t)\int_{t_0}^{t}(I - P)X^{-1}(s)b(s)\,ds \qquad \text{(A.12)}$$

$$= X(t)[(I - P)X^{-1}(t_0)x(t_0) + \int_{t_0}^{t}(I - P)X^{-1}(s)b(s)\,ds]$$

is also bounded. However, the exponential dichotomy

$$\|X(t)(I - P)X^{-1}(s)\| \leq Le^{-\beta(s-t)} \qquad \text{for } s \geq t \geq t_0,$$

implies that

$$|X(t_0)(I - P)X^{-1}(t_0)x(t_0)| =$$
$$= |X(t_0)(I - P)X^{-1}(t)X(t)(I - P)X^{-1}(t_0)x(t_0)|$$
$$\leq Le^{-\beta(t-t_0)}|X(t)(I - P)X^{-1}(t_0)x(t_0)|.$$

Hence,

$$|X(t)(I - P)X^{-1}(t_0)x(t_0)| \geq |X(t_0)(I - P)X^{-1}(t_0)x(t_0)|L^{-1}e^{\beta(t-t_0)}$$

is not bounded and since

$$|X(t)(I - P)X^{-1}(t_0)x(t_0)| \leq \|X(t)\|\,|(I - P)X^{-1}(t_0)x(t_0)|,$$

it follows that the matrix $X(t)$ is not bounded. Therefore, from (A.12) we get that the solution $x(t)$ only can be bounded if it holds

$$(I - P)X^{-1}(t_0)x(t_0) + \int_{t_0}^{\infty}(I - P)X^{-1}(s)b(s)\,ds = 0.$$

Consequently, from (A.11) we obtain that if $x^{+}(t)$ is a bounded solution of (A.9) then (A.10) holds.

Conversely, to verify that $x^{+}(t)$ given in (A.10) is a bounded solution of (A.9) it suffices to see that

$$L^{+}b(t) = \int_{t_0}^{t}X(t)X^{-1}(s)\mathscr{P}(s)b(s)\,ds$$

$$- \int_{t}^{\infty}X(t)X^{-1}(t_0)(I - \mathscr{P}(s))b(s)\,ds$$

is a particular bounded solution of the above complete linear equation. Indeed, by the Leibniz rule

$$\frac{d}{dt} L^+ b(t) = A(t) L^+ b(t) + b(t),$$

and using the dichotomy estimates

$$|L^+ b(t)| \leq \int_{t_0}^{t} K e^{-\alpha(t-s)} |b(s)| \, ds + \int_{t}^{\infty} L e^{-\beta(s-t)} |b(s)| \, ds \leq \tilde{K} + \tilde{L}$$

for all $t \geq t_0$, where the constant $\tilde{K}$ and $\tilde{L}$ not depend on t. $\qquad\square$

In the same way, a similar result to the negative bounded solutions of the complete linear equation follows.

Lemma A.18. *Assume that the linear equation* (A.5) *has an exponential dichotomy on* $J = (-\infty, t_0]$. *Let* $b \in C_b^0(J, \mathbb{R}^n)$. *Then,* $x^-(t)$ *is a positive bounded solution of* (A.9) *if and only if*

$$
\begin{aligned}
x^-(t) = {} & X(t) X^{-1}(t_0)(I - \mathcal{P}(t_0)) x^-(t_0) \\
& + \int_{-\infty}^{t} X(t) X^{-1}(s) \mathcal{P}(s) b(s) \, ds \\
& - \int_{t}^{t_0} X(t) X^{-1}(s)(I - \mathcal{P}(s)) b(s) \, ds.
\end{aligned}
\tag{A.13}
$$

Remark A.19. *Notice that the functions given in* (A.10) *and* (A.13) *are both solutions of* (A.9) *for any continuous function* $b(t)$ *on* $[t_0, \infty)$ *and* $(-\infty, t_0]$, *respectively. On the other hand, to prove that both functions are bounded solutions of this equation we need to use that* $b(t)$ *is also bounded.*

The main application of exponential dichotomies is in the context of the variational equations along certain bounded solutions of a non-linear differential equation. For this reason, the following section focuses on study the exponential dichotomy for this type of equations.

A.5 Dichotomy for the variational equation

Linear differential equations of the form of (A.5) arise in the study of the flow around of a solution $\gamma = \{p(t) : t \in J\}$ of a non-linear equation $\dot{x} = f(x)$. It is well know that the differential of the flow along $p(t)$ is given by the solution of the initial matrix problem:

$$\dot{X} = Df(p(t))X, \qquad X(0) = \mathrm{Id}.$$

This solution is a fundamental matrix of the *variational equation*

$$\dot{x} = Df(p(t))x, \qquad x \in \mathbb{R}^n, \ t \in J.$$

Special orbits of $\dot{x} = f(x)$ are the equilibrium points, the periodic orbits and the (homo)heteroclinic connections, all of them defined on $J = (-\infty, \infty)$. Dichotomy of equilibrium points correspond to notion of hyperbolicity as we see in §A.1. We are interested in studying the exponential dichotomy of the (homo)heteroclinic solutions to apply the results in the next chapter. Dichotomy of periodic orbits and non-stationary hyperbolic solutions are studied in Palmer (1984, 1996, 2000).

Let p_+ and p_- be a pair of hyperbolic equilibrium points of a non-linear equation $\dot{x} = f(x)$ where $x \in \mathbb{R}^n$ and f is a regular enough vector field. Assume that it has a orbit $\gamma = \{p(t) : t \in (-\infty, \infty)\}$ connecting p_+ and p_-. That is,

$$\lim_{t \to \infty} p(t) = p_+ \quad \text{and} \quad \lim_{t \to -\infty} p(t) = p_-.$$

The trajectory γ is called *heteroclinic* orbit if $p_+ \neq p_-$ and *homoclinic* orbit if $p_+ = p_-$. Consider the variational equation

$$\dot{x} = Df(p(t))x, \qquad x \in \mathbb{R}^n, \ t \in (-\infty, \infty). \tag{A.14}$$

According to Proposition A.11, equation (A.14) has the same exponential dichotomy than $\dot{x} = Df(p_+)x$ on $[t_0, \infty)$. Analogously, (A.14) has the same exponential dichotomy of $\dot{x} = Df(p_-)x$ on $(-\infty, t_0]$. That is, if the stable (resp. unstable) subspace of $\dot{x} = Df(p_+)x$ (resp. $\dot{x} = Df(p_-)x$) has dimension k then (A.14) has an exponential dichotomy on $[t_0, \infty)$ (resp. $(-\infty, t_0]$) with stable subspace $E_{t_0}^s$ (resp. unstable subspace $E_{t_0}^u$) with dimension k. In fact we have the following result:

Proposition A.20. *Let $p(t)$ be a solution of the equation $\dot{x} = f(x)$ parametrizing an orbit on the stable (resp. unstable) manifold of an equilibrium point p. Hence the variational equation $\dot{x} = Df(p(t))x$ has an exponential dichotomy on $[t_0, \infty)$ (resp. $(-\infty, t_0]$). Moreover,*

$$\mathcal{R}(\mathscr{P}(t_0)) = T_{p(t_0)}W^s(p) \quad (\textit{resp. } \mathcal{N}(\mathscr{P}(t_0)) = T_{p(t_0)}W^u(p)).$$

Notice that in the case of a (homo)heteroclinic orbit, there is no exponential dichotomy on $(-\infty, \infty)$. Indeed, since $f(p(t))$ is a solution of (A.14), the stable subspace $E_{t_0}^s$ and the unstable subspace $E_{t_0}^u$ have the subspace generated by the vector $f(p(t_0))$ which we denote by $\operatorname{span}\{f(p(t_0))\}$. Thus, according to Lemma A.9, the variational equation (A.14) does not have exponential dichotomy on $(-\infty, \infty)$.

As we have already noticed, the number of linearly independent forward (resp. backward) bounded solutions of the variational equation (A.14) is given by the dimension of the stable (resp. unstable) subspace of the equation $\dot{x} = Df(p_+)x$ (resp. $\dot{x} = Df(p_-)x$). That is, such number coincides with the dimension of $W^s(p_+)$ (resp. $W^u(p_-)$). Therefore, taking into account that

$$E_{t_0}^s = T_{p(t_0)}W^s(p_+) \quad \text{and} \quad E_{t_0}^u = T_{p(t_0)}W^u(p_-),$$

we can conclude, from Proposition A.15, the following result.

Proposition A.21. *If $p(t)$ is a (homo)heteroclinic solution connecting two equilibrium points p_+ and p_- then the number of linearly independent bounded solutions of the adjoint variational equation*

$$\dot{w} = -Df(p(t))^*w, \qquad w \in \mathbb{R}^n, \ t \in (-\infty, \infty)$$

is the codimension of $T_{p(t_0)}W^s(p_+) + T_{p(t_0)}W^u(p_-)$, that is,

$$\begin{aligned} c = n &- \dim W^s(p_+) - \dim W^u(p_-) \\ &+ \dim T_{p(t_0)}W^s(p_+) \cap T_{p(t_0)}W^u(p_-). \end{aligned}$$

The above number c is called *codimension* of the (homo)heteroclinic orbit $\gamma = \{p(t) : t \in (-\infty, \infty)\}$ (or of the tangency between the corresponding invariant manifolds of p_+ and p_-). Observe that

$$c = d - (s_+ - s_-) \geqslant 0$$

where $s_\pm$ are the stability indices of $p_\pm$ and

$$d = \dim T_{p(t_0)}W^s(p_+) \cap T_{p(t_0)}W^u(p_-) \geqslant 1.$$

The number d is called *dimension* of tangency along the connection γ. If γ is a homoclinic orbit, i.e., $p_+ = p_-$, then the codimension of γ coincides with the dimension, i.e., $c = d$. If $c = 0$ then γ must be a heteroclinic orbit and the connection between the invariant manifold of p_+ and p_- is transversal. Thus, persistent under perturbations. When $c > 0$ indicates that we have a bifurcation as in the case of homoclinic orbits. If $s_- = s_+$ the heteroclinic connection is said to be *equidimensional*. Otherwise, γ is said to be *heterodimensional*.

Definition A.22. *A (homo)heteroclinic orbit γ is non-degenerate if*

$$\dim T_p W^s(p_+) \cap T_p W^u(p_-) = 1,$$

with $p \in \gamma$. Otherwise γ is said to be degenerate.

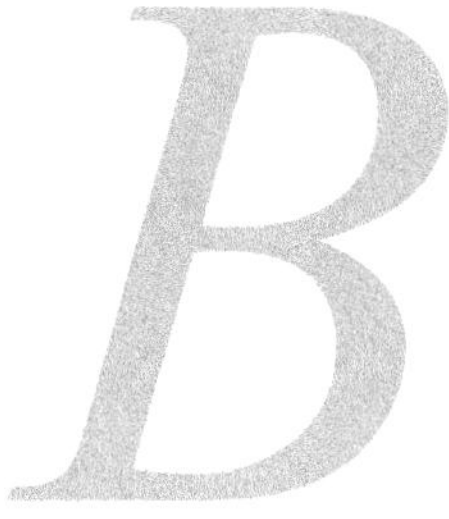

B   Continuation of connections

B.1   Introduction

Let $\dot{x} = f(x)$ be a nonlinear equation, where $x \in \mathbb{R}^n$ and f is a regular enough vector field. Assume that it has an orbit $\gamma = \{p(t) : t \in \mathbb{R}\}$ connecting two hyperbolic equilibrium points p_+ and p_-. Recall that if $p_+ = p_-$, γ is said homoclinic and otherwise heteroclinic. Consider a family

$$\dot{x} = f(x) + g(\lambda, x) \tag{B.1}$$

with $\lambda \in \mathbb{R}^k$ and g regular enough, such that $g(0, x) = 0$. For any λ small enough, family (B.1) has hyperbolic equilibrium points $p_+(\lambda)$ and $p_-(\lambda)$, continuation of p_+ and p_-, respectively, and the stability index is preserved. In order to study the persistence of the γ for λ small enough we introduce the change of variables $x = z + p(t)$ in (B.1) to obtain

$$\dot{z} = Df(p(t))z + b(\lambda, t, z) \tag{B.2}$$

where

$$b(\lambda, t, z) = f(p(t) + z) - f(p(t)) - Df(p(t))z + g(\lambda, p(t) + z).$$

Notice that $b(0, t, 0) = D_z b(0, t, 0) = 0$ for all $t \in \mathbb{R}$.

Persistence of (homo)heteroclinic orbits in (B.1) implies the existence of bounded solutions for (B.2) which, in turn, implies the existence of bounded solutions for a equation as

$$\dot{z} = Df(p(t))z + b(t) \tag{B.3}$$

where b belongs to the space $C_b^0(\mathbb{R}, \mathbb{R}^n)$.

Since f is regular ($f \in C^r$, $r \geq 1$) and $p(t)$ is a bounded solution, $Df(p(t))$ is a bounded and continuous matrix on $(-\infty, \infty)$. On the other hand, according to Proposition A.20 the variational equation $\dot{z} = Df(p(t))z$ has an exponential dichotomy on $(-\infty, t_0]$ and $[t_0, \infty)$. Thus, we are in the assumptions of Theorem A.16 and consequently, (B.3) has a bounded solution if and only if

$$\int_{-\infty}^{\infty} <w(t), b(t)> \, dt = 0$$

for all bounded solution $w(t)$ of the adjoint variational equation.

According to Proposition A.21, the adjoint variational equation

$$\dot{w} = -Df(p(t))^* w$$

has c linearly independent bounded solutions w_i where c is the codimension of γ. Then the persistence of the (homo)heteroclinic orbit γ requires the fulfillment of the c conditions

$$\int_{-\infty}^{\infty} \langle w_i(t), b(t) \rangle \, dt = 0 \quad \text{for } i = 1, \dots, c.$$

The question is the sufficiency of such conditions. When $c = 1$ the sufficiency could be followed from Chow, Hale, and Mallet-Paret (1980). In general, for $c \geq 1$, the techniques to be used follow the first steps of the Lin's method in X.-B. Lin (1990) and Sandstede (1993).

B.2   Bifurcation equation

The persistence of the (homo)heteroclinic connection as the continuation of $\gamma = \{p(t) : t \in (-\infty, \infty)\}$ is given by certain bifurcation equation which we will obtain as a consequence of the Lin's method. For $\|\lambda\|$ small enough, one has to look for solutions $p_\lambda^+(\cdot)$ and $p_\lambda^-(\cdot)$ of (B.1), contained in the stable and unstable invariant manifolds of the equilibrium points $p_+(\lambda)$ and $p_-(\lambda)$, respectively (see Figure B.1). Initial values $p_\lambda^\pm(t_0)$ will belong to a section Σ_{t_0} transverse to

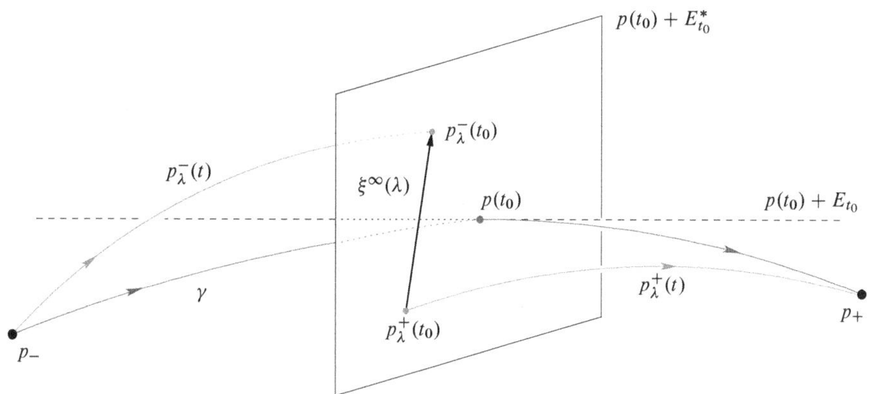

Figure B.1: *Non-degenerate heteroclinic orbit in $\mathbb{R}^3$ where the 1-dimensional manifolds coincide. In this case, $E_{t_0}^s = E_{t_0}^u = E_{t_0}$ (1-dimensional), $E_{t_0}^{s*} = E_{t_0}^{u*} = E_{t_0}^*$ (2-dimensional) and $\Sigma_{t_0} = p(t_0) + E_{t_0}^*$. For simplicity, we have assume that the perturbation satisfies $p_-(\lambda) = p_-$ and $p_+(\lambda) = p_+$ for all λ.*

the (homo)heteroclinic orbit γ. Namely $\Sigma_{t_0} = p(t_0) + \{f(p(t_0))\}^\perp$. Moreover the condition

$$\xi^\infty(\lambda) = p_{\bar\lambda}^-(t_0) - p_{\bar\lambda}^+(t_0) \in E_{t_0}^*$$

will be required where $E_{t_0}^* = E_{t_0}^{s*} \cap E_{t_0}^{u*} = [E_{t_0}^s + E_{t_0}^u]^\perp \subset \{f(p(t_0))\}^\perp$. Under these assumptions there will exist two unique solutions $p_{\bar\lambda}^\pm(\cdot)$ for each λ. The jump $\xi^\infty(\lambda)$ measures the displacement between the stable and unstable invariant manifolds on the section Σ_{t_0} in the direction of the subspace $E_{t_0}^*$. In the two following sections we will deduce the bifurcation equation $\xi^\infty(\lambda) = 0$ in the cases that γ is a non-degenerate and degenerate (homo)heteroclinic connection.

B.3 Non-degenerate connections

In this subsection we assume the following hypothesis: - The (homo)heteroclinic orbit $\gamma = \{p(t) : t \in (-\infty, \infty)\}$ is non-degenerate and of codimension $c \geqslant 1$.

According to Proposition A.21, the number of linear independent bounded solutions of the adjoint variational equation is equal to c. At the same time this number coincides with the dimension of $E_{t_0}^* = [E_{t_0}^s + E_{t_0}^u]^\perp$. Moreover, since γ is non-degenerate

$$E_{t_0} = E_{t_0}^s \cap E_{t_0}^u = T_{p(t_0)}W^s(p_+) \cap T_{p(t_0)}W^u(p_-)$$

has dimension one. Namely, it is the space generated by the vector $\dot{p}(t_0) = f(p(t_0))$. Thus,

$$c = n - \dim W^s(p_+) - \dim W^u(p_-) + 1.$$

We introduce $W_{t_0}^{\pm}$ as the orthogonal complement of E_{t_0} in the tangent space to the stable and unstable manifold of p_+ and p_- respectively. That is,

$$E_{t_0}^s = E_{t_0} \perp W_{t_0}^+ \quad \text{and} \quad E_{t_0}^u = E_{t_0} \perp W_{t_0}^-.$$

Then

$$\mathbb{R}^n = \text{span}\{f(p(t_0))\} \oplus W_{t_0}^+ \oplus W_{t_0}^- \oplus E_{t_0}^*. \tag{B.4}$$

Finally we take the transversal section to γ at $p(t_0)$ given by

$$\Sigma_{t_0} = p(t_0) + \{f(p(t_0))\}^{\perp} = p(t_0) + [W_{t_0}^+ \oplus W_{t_0}^- \oplus E_{t_0}^*].$$

Figure B.2 shows the transversal section Σ_{t_0}. The curves W_λ^s and W_λ^u are, respectively, $W^s(p_+(\lambda)) \cap \Sigma_{t_0}$ and $W^u(p_-(\lambda)) \cap \Sigma_{t_0}$. A priori the curve W_λ^s does not met W_λ^u. However, the projection along the direction $E_{t_0}^*$ of both curves on $p(t_0) + [W_{t_0}^- \oplus W_{t_0}^+]$ have a unique transversal intersection point q_λ. Now, $q_\lambda + E_{t_0}^*$ meets, respectively, W_λ^s and W_λ^u at $p_\lambda^+(t_0)$ and $p_\lambda^-(t_0)$. These two points define the vector $\xi^\infty(\lambda)$. The persistence of the connection holds if $\xi^\infty(\lambda) = 0$ which provides a set of $c = \dim E_{t_0}^*$ conditions.

The proof of the next result can be found in Sandstede (1993, Lem. 3.3) and Knobloch (2004, Lem. 2.1.2). Namely, in Knobloch (2004) only the first item is proved and, moreover, the proof is developed for the degenerate case although the non-degenerate one follows in a similar manner. The second item is proved in Sandstede (1993) for the non-degenerate case. We include here a complete and simplified proof of this result.

Lemma B.1. *There is $\delta > 0$ such that for every $\lambda \in \mathbb{R}^k$ with $|\lambda| < \delta$,*

1. *there exists a unique pair of solutions $p_\lambda^+(t)$ and $p_\lambda^-(t)$ of (B.1) parameterizing orbits on $W^s(p_+(\lambda))$ and $W^u(p_-(\lambda))$, respectively, such that $p_\lambda^{\pm}(t_0) \in \Sigma_{t_0}$ and*

$$\xi^\infty(\lambda) = p_\lambda^-(t_0) - p_\lambda^+(t_0) \in E_{t_0}^*.$$

Writing the solutions as $p_\lambda^{\pm}(t) = p(t) + z_\lambda^{\pm}(t)$, then $z_\lambda^{\pm}(\cdot)$ are, respectively, forward and backward bounded solutions of the equation (B.2). They depend regularly on λ and the functions $z_0^{\pm}$ are identically zero.

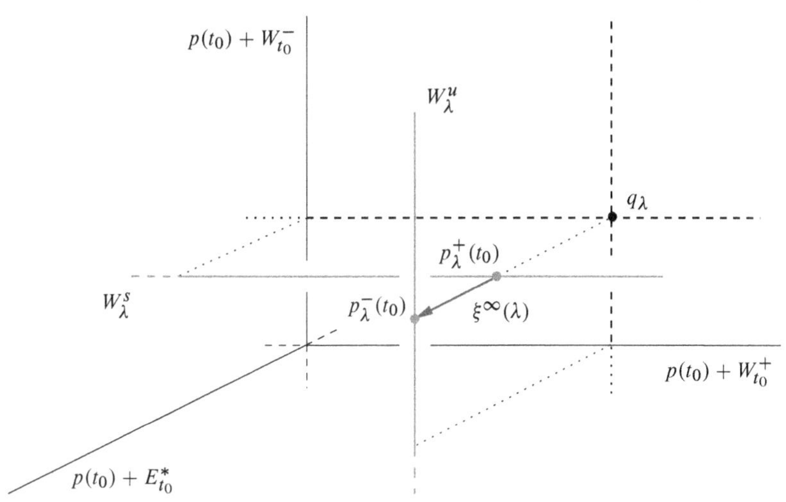

Figure B.2: *Non-degenerate (homo)heteroclinic orbit*

2. *for $\varepsilon > 0$ small enough, there exists a (homo)heteroclinic solution $p_\lambda(t)$ such that*

$$\|p_\lambda(t_0) - p(t_0)\| < \varepsilon \text{ if and only if } \xi^\infty(\lambda) = 0.$$

That is, the components $\xi_i^\infty(\lambda)$ of the vector $\xi^\infty(\lambda)$ in a basis $\{w_i : i = 1, \ldots, c\}$ of $E_{t_0}^*$ satisfy

$$\xi_i^\infty(\lambda) \equiv \int_{-\infty}^{t_0} < w_i(s), b(\lambda, s, z_\lambda^-(s)) > ds$$
$$+ \int_{t_0}^{\infty} < w_i(s), b(\lambda, s, z_\lambda^+(s)) > ds = 0$$

being $w_i(s) = X^{-1}(s)^* X(t_0)^* w_i$ for $i = 1, \ldots, c$ bounded linearly independent solutions of the adjoint variational equation.

Proof. The solutions $p_\lambda(t)$ of (B.1) can be written as $p_\lambda(t) = p(t) + z_\lambda(t)$ where $z_\lambda(t)$ is a solution of (B.2). Let $Y_{t_0} = W_{t_0}^+ \oplus W_{t_0}^- \oplus E_{t_0}^*$. Assuming $p_\lambda(t_0) \in \Sigma_{t_0}$ then $z_\lambda(t_0) \in Y_{t_0}$.

In order to get that $p_\lambda(t)$ parametrizes an orbit in $W^s(p_+(\lambda))$ (resp. $W^u(p_-(\lambda))$), the function $z_\lambda(t)$ have to be a positive (resp. negative) bounded solution of (B.2). Now if we assume that $z^+(t)$ and $z^-(t)$ are a pair of positive and negative bounded solutions of (B.2) respectively, then $b(\cdot, z^\pm(\cdot), \lambda) \in$

$C_b^0(J_\pm, \mathbb{R}^n)$ where $J_+ = [t_0, \infty)$ and $J_- = (-\infty, t_0]$. Thus, according to Lemma A.17 and Lemma A.18 it must be met that

$$
\begin{aligned}
z^+(t) = {}& X(t)X^{-1}(t_0)\mathscr{P}_+(t_0)z^+(t_0) \\
& + \int_{t_0}^t X(t)X^{-1}(s)\mathscr{P}_+(s)b(s, z^+(s), \lambda)\, ds \\
& - \int_t^\infty X(t)X^{-1}(s)(I - \mathscr{P}_+(s))b(s, z^+(s), \lambda)\, ds
\end{aligned}
\tag{B.5}
$$

and

$$
\begin{aligned}
z^-(t) = {}& X(t)X^{-1}(t_0)(I - \mathscr{P}_-(t_0))z^-(t_0) \\
& + \int_{-\infty}^t X(t)X^{-1}(s)\mathscr{P}_-(s)b(s, z^-(s), \lambda)\, ds \\
& - \int_t^{t_0} X(t)X^{-1}(s)(I - \mathscr{P}_-(s))b(s, z^-(s), \lambda)\, ds
\end{aligned}
\tag{B.6}
$$

where $\mathscr{P}_+(t_0)$ and $I - \mathscr{P}_-(t_0)$ are the corresponding projection matrix on the stable space $E_{t_0}^s = T_{p(t_0)}W^s(p_+)$ and unstable space $E_{t_0}^u = T_{p(t_0)}W^u(p_-)$, respectively.

Conversely, according to Remark A.19, the solutions $z^+(t)$ and $z^-(t)$ of the integral equations (B.5) and (B.6) are both solutions of (B.2), but not necessarily bounded. The existence of positive and negative bounded solutions of (B.5) and (B.6) will be proved as an application of Implicit Function Theorem.

Let $\eta^+ = \mathscr{P}_+(t_0)z^+(t_0) \in W_{t_0}^+$ and $\eta^- = (I - \mathscr{P}_-(t_0))z^-(t_0) \in W_{t_0}^-$. Equations (B.5) and (B.6) can be written in the form

$$
z^\pm = \mathscr{H}^\pm(z^\pm, \eta^\pm, \lambda)
\tag{B.7}
$$

where $\mathscr{H}^\pm : C_b^0(J_\pm, \mathbb{R}^n) \times W_{t_0}^\pm \times \mathbb{R}^k \to C_b^0(J_\pm, \mathbb{R}^n)$. In order to apply the Implicit Function Theorem to the equation (B.7) notice first that $\mathscr{H}^\pm(0, 0, 0) = 0$. On the other hand,

$$
D_z\mathscr{H}^\pm(z^\pm, \eta^\pm, \lambda) : C_b^0(J_\pm, \mathbb{R}^n) \to C_b^0(J_\pm, \mathbb{R}^n)
$$

is the null function for $z^\pm = \eta^\pm = \lambda = 0$. Indeed, for any $h \in C_b^0(J_+, \mathbb{R}^n)$ it holds that

$$
\begin{aligned}
D_z\mathscr{H}^+(z, \eta, \lambda)h(t) = {}& \int_{t_0}^t X(t)X^{-1}(s)\mathscr{P}_+(s)D_z b(s, z(s), \lambda)h(s)\, ds \\
& - \int_t^\infty X(t)X^{-1}(s)(I - \mathscr{P}_+(s))D_z b(s, z(s), \lambda)h(s)\, ds.
\end{aligned}
$$

Since $D_z b(s,0,0) = 0$ then $D_z \mathcal{H}^+(0,0,0) = 0$. Similarly it follows that $D_z \mathcal{H}^-(0,0,0) = 0$. Therefore, there exists $\delta_\pm > 0$ such that for every $\eta^\pm \in W_{t_0}^\pm$ and $\lambda \in \mathbb{R}^k$ with $|\eta^\pm|, |\lambda| < \delta_\pm$ there is a unique $z^\pm(\eta^\pm, \lambda) \in C_b^0(J_\pm, \mathbb{R}^n)$ so that

$$z^\pm(\eta^\pm, \lambda) = \mathcal{H}^\pm(z^\pm(\eta^\pm, \lambda), \eta^\pm, \lambda) \quad \text{and} \quad z^\pm(0,0) = 0.$$

Now, consider the condition $z^-(\eta^-, \lambda,)(t_0) - z^+(\eta^+, \lambda)(t_0) \in E_{t_0}^*$. Since $Y_{t_0} = W_{t_0}^+ \oplus W_{t_0}^- \oplus E_{t_0}^*$, we can write

$$z^\pm(\eta^\pm, \lambda)(t_0) = \eta^\pm + w^\mp(\eta^\pm, \lambda) + \varrho^\pm(\eta^\pm, \lambda)$$

where $w^\mp(\eta^\pm, \lambda) \in W_{t_0}^\mp$ and $\varrho^\pm(\eta^\pm, \lambda) \in E_{t_0}^*$. From (B.5) and (B.6), and having into account that

$$X(t)X^{-1}(s)\mathscr{P}_\pm(s) = \mathscr{P}_\pm(t)X(t)X^{-1}(s),$$

it follows that $z^-(\eta^-, \lambda)(t_0)$ and $z^+(\eta^+, \lambda)(t_0)$ are, respectively,

$$\eta^- + \mathscr{P}_-(t_0) \int_{-\infty}^{t_0} X(t_0)X^{-1}(s)b(s, z^-(\eta^-, \lambda)(s), \lambda)\, ds,$$
$$\eta^+ - (I - \mathscr{P}_+(t_0)) \int_{t_0}^{\infty} X(t_0)X^{-1}(s)b(s, z^+(\eta^+, \lambda)(s), \lambda)\, ds. \tag{B.8}$$

Thus $w^+(\eta^-, \lambda) + \varrho^-(\eta^-, \lambda)$ and $w^-(\eta^+, \lambda) + \varrho^+(\eta^+, \lambda)$ are, respectively,

$$\mathscr{P}_-(t_0) \int_{-\infty}^{t_0} X(t_0)X^{-1}(s)b(s, z^-(\eta^-, \lambda)(s), \lambda)\, ds,$$
$$- (I - \mathscr{P}_+(t_0)) \int_{t_0}^{\infty} X(t_0)X^{-1}(s)b(s, z^+(\eta^+, \lambda)(s), \lambda)\, ds. \tag{B.9}$$

Recall that $z^\pm(0,0) = 0$ and hence $w^\pm(0,0) = \varrho^\pm(0,0) = 0$. On the other hand, applying that $D_z b(t,0,0) = 0$ in (B.9) we get that

$$D_{\eta^\mp}\varrho^\mp(0,0) = D_\lambda \varrho^\mp(0,0) = 0,$$
$$D_{\eta^\mp}w^\pm(0,0) = D_\lambda w^\pm(0,0) = 0. \tag{B.10}$$

The condition $z^-(\eta^-, \lambda)(t_0) - z^+(\eta^+, \lambda)(t_0) \in E_{t_0}^*$ is equivalent to the system of two equations $\eta^\pm - w^\pm(\eta^\mp, \lambda) = 0$ which we write in the form $F(\eta, \lambda) = 0$ where $\eta = (\eta^+, \eta^-) \in W_{t_0}^+ \times W_{t_0}^-$. We have that $F(0,0) = 0$ and according to (B.10) it follows that $D_\eta F(0,0) = I$. Hence, applying again the

Implicit Function Theorem we get η as a function of λ. We conclude that there exists $\delta > 0$ ($\delta < \delta_\pm$) such that for any $\lambda \in \mathbb{R}^k$ with $|\lambda| < \delta$ there is a unique $\eta^\pm(\lambda) \in W_{t_0}^\pm$ so that

$$\eta^\pm(\lambda) - w^\pm(\eta^\mp(\lambda), \lambda) = 0 \quad \text{and} \quad \eta^\pm(0) = 0.$$

Now consider the functions $z_\lambda^\pm(t) = z^\pm(\eta^\pm(\lambda), \lambda)(t)$. The uniqueness, boundedness and regularity respect to λ of $z_\lambda^\pm(t)$, and thus of $p_\lambda^\pm(t) = p(t) + z_\lambda^\pm(t)$, are following from the Implicit Function Theorem. Also we have that $z_0^\pm = 0$ and since for $|\lambda|$ small enough $z_\lambda^\pm$ is close to $z_0^\pm = 0$ we get that $\sup_{t \in J_\pm} |p_\lambda^\pm(t) - p(t)|$ is arbitrarily small. This means that the orbits parameterized by $p_\lambda^\pm(t)$ are close to the orbit γ which is given by $p(t)$. So, together the hyperbolicity of the equilibria $p_\pm(\lambda)$ we get that $\lim_{t \to \pm\infty} p_\lambda^\pm(t) = p_\pm(\lambda)$. Thus, for each λ with $|\lambda| < \delta$, the solutions $p_\lambda^\pm(t)$ parameterize, respectively, orbits in the stable manifold $W^s(p_+(\lambda))$ and in the unstable manifold $W^u(p_-(\lambda))$ such that $p_\lambda^\pm(t_0) \in p(t_0) + Y_{t_0} = \Sigma_{t_0}$ and

$$\xi^\infty(\lambda) = p_\lambda^-(t_0) - p_\lambda^+(t_0) = z_\lambda^-(t_0) - z_\lambda^+(t_0) \in E_{t_0}^*.$$

This proves the first item of the lemma.

In order to prove the second item notice that if $\xi^\infty(\lambda) = 0$ then a (homo)heteroclinic orbit of (B.1) is given by

$$p_\lambda(t) = \begin{cases} p_\lambda^-(t) & \text{para } t \leq t_0, \\ p_\lambda^+(t) & \text{para } t \geq t_0. \end{cases}$$

where $p_\lambda^\pm(t)$ are the solutions in the first item. On the other hand, if $p_\lambda(t)$ is a solution parametrizing a (homo)heteroclinic connection such that $p_\lambda(t_0) \in \Sigma_{t_0}$ with $|\lambda|$ and $|p_\lambda(t_0) - p(t_0)|$ small enough, its restriction to the intervals $J_- = (-\infty, t_0]$ and $J_+ = [t_0, \infty)$ define a pair of solutions $p_\lambda^\pm(t)$ in the assumption of the first item. That is, $p_\lambda^\pm(t_0) \in \Sigma_{t_0}$ and $p_\lambda^-(t_0) - p_\lambda^+(t_0) = 0 \in E_{t_0}^*$. This makes obvious the reciprocal implication.

To conclude the proof of the second item we will consider a base $\{w_i : i = 1 \ldots c\}$ of $E_{t_0}^* = E_{t_0}^{s*} \cap E_{t_0}^{u*} = [E_{t_0}^s + E_{t_0}^u]^\perp$. Then

$$\xi^\infty(\lambda) = \sum_{i=1}^{c} < w_i, \xi^\infty(\lambda) > w_i.$$

Form (B.8) and having into account that $< w_i, \eta^\pm > = 0$, it follows

$$\xi_i^\infty(\lambda) \overset{\text{def}}{=} < w_i, \xi^\infty(\lambda) >$$

$$= < w_i, \mathscr{P}_-(t_0) \int_{-\infty}^{t_0} X(t_0)X^{-1}(s)b(s, z_\lambda^-(s), \lambda)\, ds >$$

$$+ < w_i, (I - \mathscr{P}_+(t_0)) \int_{t_0}^{\infty} X(t_0)X^{-1}(s)b(s, z_\lambda^+(s), \lambda)\, ds >$$

Thus, $\xi_i^\infty(\lambda)$ is given by

$$\int_{-\infty}^{t_0} < [\mathscr{P}_-(t_0)X(t_0)X^{-1}(s)]^* w_i, b(s, z_\lambda^-(s), \lambda)\, ds > \qquad \text{(B.11)}$$

$$+ \int_{t_0}^{\infty} < [(I - \mathscr{P}_+(t_0))X(t_0)X^{-1}(s)]^* w_i, b(s, z_\lambda^+(s), \lambda)\, ds > .$$

Thus, since $\mathscr{P}_-(t_0)^* : \mathbb{R}^n \to E_{t_0}^{u*}$ and $I - \mathscr{P}_+(t_0)^* : \mathbb{R}^n \to E_{t_0}^{s*}$ we get that

$$[\mathscr{P}_-(t_0)X(t_0)X^{-1}(s)]^* w_i = [(I - \mathscr{P}_+(t_0))X(t_0)X^{-1}(s)]^* w_i = w_i(s).$$

Substituting in (B.11) we obtain that

$$\xi_i^\infty(\lambda) \equiv \int_{-\infty}^{t_0} < w_i(s), b(s, z_\lambda^-(s), \lambda) > \, ds$$

$$+ \int_{t_0}^{\infty} < w_i(s), b(s, z_\lambda^+(s), \lambda) > \, ds = 0.$$

This concludes the second item and proves of the lemma. $\qquad \square$

According to the above statement the persistence of a (homo)heteroclinic orbit follows from the analysis of the bifurcation equation $\xi^\infty(\lambda) = 0$. The existence of non-zero parameter values $\lambda \in \mathbb{R}^k$ such that $\xi^\infty(\lambda) = 0$ follows from the Implicit Function Theorem when $D_\lambda \xi^\infty(0)$ has rank $c < k$. Thus, the following result follows:

Theorem B.2. *Let* $\xi^\infty(\lambda) = 0$, *with* $\lambda \in \mathbb{R}^k$, *be the bifurcation equation of (B.1). If* $k > c$ *and* $\text{rank } D_\lambda \xi^\infty(0) - c$, *then (B.1) has a (homo)heteroclinic orbit for each parameter value* λ *on a regular manifold of dimension* $k - c$ *with tangent subspace at* $\lambda = 0$ *given by the solutions of the system*

$$\sum_{j=1}^{k} \xi_{ij}^\infty \lambda_j - 0 \qquad i = 1, \ldots, c$$

where

$$\xi_{ij}^{\infty} \equiv \frac{\partial \xi_i^{\infty}}{\partial \lambda_j}(0) = \int_{-\infty}^{\infty} < w_i(s), D_{\lambda_j} g(0, p(s)) > ds$$

for $i = 1, \ldots, c$ and $j = 1, \ldots, k$.

Remark B.3. *Note that, when $k \leqslant c$, $\lambda = 0$ is the unique value of $\lambda \in \mathbb{R}^k$ for which there is a (homo)heteroclinic orbit*

$$\gamma_\lambda = \{ p_\lambda(t) : \ \dot{p}_\lambda(t) = f(p_\lambda(t)) + g(\lambda, p_\lambda(t)), \ t \in \mathbb{R} \}$$

such that $\sup_{t \in \mathbb{R}} \| p_\lambda(t) - p(t) \|$ is small enough. If $k > c$ the (homo)-heteroclinic connection persists for parameter values on a manifold of codimension c. We say that there is (homo)heteroclinic bifurcation of a non-degenerate orbit at $\lambda = 0$ which is of codimension c.

B.3.1 Heteroclinic connections

In the case that γ is a heteroclinic orbit ($p_+ \neq p_-$), under the assumption (B.3) we have that the codimension of γ is $c = s_- - s_+ + 1$ where $s_\pm$ are the stability indices of $p_\pm$. Since $c \geqslant 0$ implies that $s_- \geqslant s_+ - 1$. and we have the following remark:

Remark B.4. *Every non-degenerate heteroclinic orbit to hyperbolic equilibrium point of stability indices $s_\pm$ with $s_- \geqslant s_+ - 1$ persists under k-parameter perturbations for value of the parameter on a manifold of dimension $k - (s_- - s_+ + 1)$.*

The bifurcation equation is $\xi^\infty(\lambda) = 0$ where $\xi^\infty : \mathbb{R}^k \to \mathbb{R}^c$ with $c = s_- - s_+ + 1 \leqslant k$. Let $w_i(s)$, $i = 1, \ldots, c$ be linearly independent bounded solutions of the adjoint variational equation $\dot{w} = -Df(p(t))^* w$. It follows from Theorem B.2 that under the generic condition

$$\text{range } D\xi^\infty(0) = c, \qquad D\xi^\infty(0) = (\xi_{ij}^\infty)_{i=1,\ldots,c, \ j=1,\ldots,k}$$

where

$$\xi_{ij}^\infty = \int_{-\infty}^{\infty} < w_i(s), D_{\lambda_j} g(p(s), 0) > ds$$

the tangent space at $\lambda = 0$ to the $(k - c)$-dimensional manifold of parameters for which there exist heteroclinic orbits continuation of the primary γ is

$$\xi_{i1}^\infty \lambda_1 + \cdots + \xi_{ik}^\infty \lambda_k = 0 \qquad \text{para } i = 1, \ldots, c.$$

B.3.2 Homoclinic connections

In the case that γ is a homoclinic orbit $(p_+ = p_- = p)$ then

$$\dim E_{t_0}^* = \dim T_{p(t_0)} W^s(p) \cap T_{p(t_0)} W^u(p) = \dim E_{t_0}.$$

Under the assumption (B.3) we have that $\dim E_{t_0} = 1$ and thus the bifurcation is of codimension $c = 1$. This means the following:

Remark B.5. *Every non-degenerate homoclinic orbit to a hyperbolic equilibrium point persists under k-parameter perturbations for value of the parameter on a hypersurface of dimension $k - 1$.*

In this case, the bifurcation equation is $\xi^\infty(\lambda) = 0$ where ξ^∞ is an scalar function. Let $w(s)$ be the unique bounded (linearly independent) solution of the adjoint variational equation $\dot{w} = -Df(p(t))^* w$. Under the generic condition

$$\nabla \xi^\infty(0) = (\xi^\infty_{\lambda_1}, \ldots, \xi^\infty_{\lambda_k}) \neq 0$$

where

$$\xi^\infty_{\lambda_j} = \int_{-\infty}^{\infty} < w(s), D_{\lambda_j} g(p(s), 0) > ds, \quad \text{for } j = 1, \ldots, k,$$

the tangent hyperplane at $\lambda = 0$ to the hypersurface of parameters for which there exist homoclinic orbits continuation of the primary orbit γ is $\xi^\infty_{\lambda_1} \lambda_1 + \cdots + \xi^\infty_{\lambda_k} \lambda_k = 0$.

It is worth noting that all homoclinic orbits in low dimension $(n \leqslant 3)$ are non-degenerate. However, this does not occur in dimension $n \geqslant 4$ where degenerate homoclinic orbits could appeared. In the next section we will explain how to deal with this case.

B.4 Degenerate connections

We study the persistence of connections under the hypothesis: - The (homo)heteroclinic orbit $\gamma = \{p(t) : t \in (-\infty, \infty)\}$ is degenerate and of codimension $c \geqslant 1$.

The study of the persistence of a (homo)heteroclinic orbit under the hypothesis (B.4) differs from the study of persistence under the assumption (B.3) mainly in the decomposition of the cross section Σ_{t_0}. Since the dimension of the space $E_{t_0} = T_{p(t_0)} W^s(p_+) \cap T_{p(t_0)} W^u(p_-)$ is greater than one, we introduce the subspace U_{t_0} which is the orthogonal complement to the vector field direction in E_{t_0}. That is,

$$E_{t_0} = \text{span}\{f(p(t_0))\} \perp U_{t_0}.$$

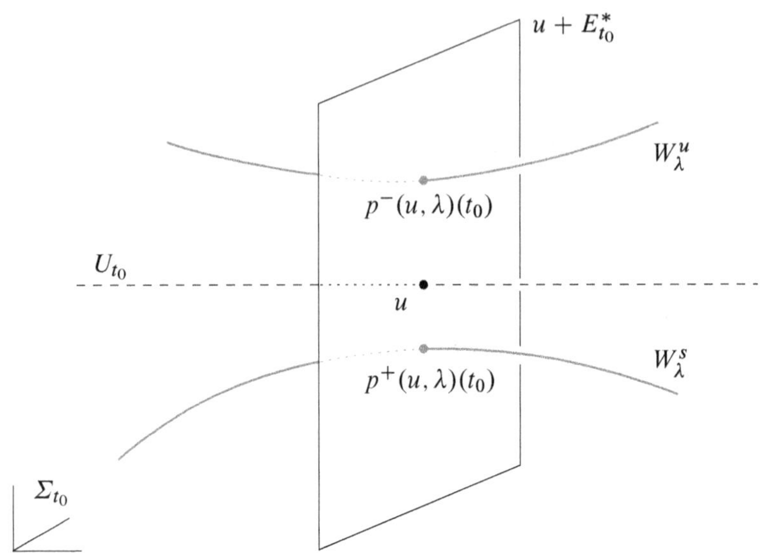

Figure B.3: *Degenerated (homo)heteroclinic connection.*

Hence, we get the following decomposition:

$$\mathbb{R}^n = \text{span}\{f(p(t_0))\} \oplus U_{t_0} \oplus W_{t_0}^+ \oplus W_{t_0}^- \oplus E_{t_0}^*. \tag{B.12}$$

As in decomposition (B.4), the spaces $W_{t_0}^{\pm}$ are the orthogonal complement of E_{t_0} in the tangent space to the stable and unstable manifold of p_+ and p_- respectively. However, the common tangent U_{t_0} provides that the dimension of

$$E_{t_0}^* = [E_{t_0}^s + E_{t_0}^u]^{\perp} = [\text{span}\{f(p(t_0))\} \oplus U_{t_0} \oplus W_{t_0}^+ \oplus W_{t_0}^-]^{\perp}$$

increases with respect to the dimension of its analog in decomposition (B.4). In short, we will consider as cross section to γ at $p(t_0)$ the following:

$$\Sigma_{t_0} = p(t_0) + \{f(p(t_0))\}^{\perp} = p(t_0) + [U_{t_0} \oplus W_{t_0}^+ \oplus W_{t_0}^- \oplus E_{t_0}^*].$$

The following result is similar to Lemma B.1. We provide a sketch of the proof. To see more details we refer to Knobloch (1997) and Knobloch (2004, Lem. 2.1.2).

Lemma B.6. *There is $\delta > 0$ such that for every $u \in U_{t_0}$, $\lambda \in \mathbb{R}^k$ with $|u|$, $|\lambda| < \delta$,*

1. *there exists a unique pair of solutions $p^+(u, \lambda)(t)$ and $p^-(u, \lambda)(t)$ of (B.1) parameterizing orbits on $W^s(p_+(\lambda))$ and $W^u(p_-(\lambda))$, respectively, such that $p^\pm(u, \lambda)(t_0) \in \Sigma_{t_0}$ and*

$$\xi^\infty(u, \lambda) = p^-(u, \lambda)(t_0) - p^+(u, \lambda)(t_0) \in E_{t_0}^*.$$

Writing the solutions as $p^\pm(u, \lambda)(t) = p(t) + z^\pm(u, \lambda)(t)$, then $z^\pm(u, \lambda)(\cdot)$ are, respectively, forward and backward bounded solutions of the equation (B.2). They depend regularly on u an λ and the functions $z^\pm(0, 0)$ are identically zero.

2. *for $\varepsilon > 0$ small enough, there exists a (homo)heteroclinic solution $p(u, \lambda)(t)$ such that*

$$|p(u, \lambda)(t_0) - p(t_0)| < \varepsilon \text{ if and only if } \xi^\infty(u, \lambda) = 0.$$

That is, the components $\xi_i^\infty(u, \lambda)$ of the vector $\xi^\infty(u, \lambda)$ in a basis $\{w_i : i = 1, \ldots, c\}$ of $E_{t_0}^*$ satisfy

$$\xi_i^\infty(u, \lambda) \equiv \int_{-\infty}^{t_0} < w_i(s), b(s, z^-(u, \lambda)(s), \lambda) > \, ds$$
$$+ \int_{t_0}^{\infty} < w_i(s), b(s, z^+(u, \lambda)(s), \lambda) > \, ds = 0.$$

being $w_i(s) = X^{-1}(s)^* X(t_0)^* w_i$ for $i = 1, \ldots, c$ bounded linearly independent solutions of the adjoint variational equation. Moreover, if R denotes the projection on U_{t_0} according to decomposition (B.12) then $R(p^\pm(u, \lambda)(t_0) - p(t_0)) = u$.

Proof. The decomposition (B.12) of the transversal section Σ_{t_0} forces to write

$$\mathscr{P}_+(t_0)z^+(t_0) = u^+ + \eta^+ \in U_{t_0} \oplus W_{t_0}^+$$

and

$$(I - \mathscr{P}_-(t_0))z^-(t_0) = u^- + \eta^- \in U_{t_0} \oplus W_{t_0}^-.$$

Then, (B.7) is written as $z^\pm = \mathscr{H}^\pm(z^\pm, u^\pm + \eta^\pm, \lambda)$ where

$$\mathscr{H}^\pm : C_a^0(J_\pm, \mathbb{R}^n) \times (U_{t_0} \oplus W_{t_0}^\pm) \times \mathbb{R}^k \to C_a^0(J_\pm, \mathbb{R}^n).$$

The rest of the proof can be reformulated in terms of the Liapunov–Schmidt reduction, for the equation

$$F(u^+ + \eta^+, u^- + \eta^-, z^+, z^-, \lambda) = 0 \qquad \text{(B.13)}$$

defines from $(U_{t_0} \oplus W_{t_0}^+) \times (U_{t_0} \oplus W_{t_0}^-) \times C_a^0(J_+, \mathbb{R}^n) \times C_a^0(J_-, \mathbb{R}^n) \times \mathbb{R}^k$
to $C_a^0(J_+, \mathbb{R}^n) \times C_a^0(J_-, \mathbb{R}^n) \times Y_{t_0}$ by means of

$$\left(z^+ - \mathcal{H}^+(u^+ + \eta^+, z^+, \lambda), \; z^- - \mathcal{H}^-(u^- + \eta^-, z^-, \lambda), \; z^-(t_0) - z^+(t_0)\right).$$

Since $F(0) = 0$ and the restriction of F to $\lambda = 0$ is a non-linear Fredholm
operator at $(u^+ + \eta^+, u^- + \eta^-, z^+, z^-) = 0$, the persistence of the (homo)het-
eroclinic connection, given by (B.13) is reduced to the bifurcation equation
$\xi^\infty(u, \lambda) = 0$. $\square$

From the above lemma, it follows that the bifurcation equation $\xi^\infty(u, \lambda) = 0$
depends on the space U_{t_0}. Because of this dependency, the analysis of persistence
of (homo)heteroclinic connections in the parameter space is not followed as in
Theorem B.2.

In Figure B.4 we represent $p(t_0) + [U_{t_0} \oplus W_{t_0}^+ \oplus W_{t_0}^-]$, where $U_{t_0} \oplus W_{t_0}^\pm$
are the tangent spaces of $W^s(p_+) \cap \Sigma_{t_0}$ and $W^u(p_-) \cap \Sigma_{t_0}$, respectively.
By W_λ^s and W_λ^u we denote the invariant manifolds $W^s(p_+(\lambda)) \cap \Sigma_{t_0}$ and
$W^s(p_-(\lambda)) \cap \Sigma_{t_0}$, which meet transversally on $p(t_0) + [U_{t_0} \oplus W_{t_0}^+ \oplus W_{t_0}^-]$
along the manifold U_λ with dim $U_\lambda = $ dim U_{t_0}. Notice that fixed $\hat{u} \in U_{t_0}$, at the
section $u = \hat{u}$ we have the same picture as in Figure B.2. In order to incorporate
the space $E_{t_0}^*$ in the picture, in Figure B.3 is represented the particular case with
$U_{t_0} = W_{t_0}^+ = W_{t_0}^-$ and dim $U_{t_0} = 1$. In this case dim $E_{t_0}^* = 2$.

Remark B.7. *Since $b(t, 0, 0) = D_z b(t, 0, 0) = 0$, from Lemma B.6 it follows
that $\xi^\infty(0, 0) = 0$ and $D_u \xi^\infty(0, 0) = 0$. This is the analytic description of
the fact that the stable and unstable manifolds of p_+ and p_-, respectively, are
tangent to the space U_{t_0}. The dimension of this space measure the degree of
degeneration, $d(\gamma)$, of the (homo)heteroclinic orbit γ:*

$$d(\gamma) = \dim T_{p(t_0)} W^s(p_+) \cap T_{p(t_0)} W^u(p_-) = 1 + \dim U_{t_0}.$$

*Now, if $(u, \lambda) \in U \times \mathbb{R}^k$ is the solution of the bifurcation equation $\xi^\infty(u, \lambda) = 0$,
the degree of degeneration of the corresponding orbit $\gamma(u, \lambda)$ follows also from
the bifurcation equation (see Knobloch (2004, Lem. 5.1.2)) as*

$$d(\gamma(u, \lambda)) = 1 + \dim \mathcal{N}(D_u \xi^\infty(u, \lambda)).$$

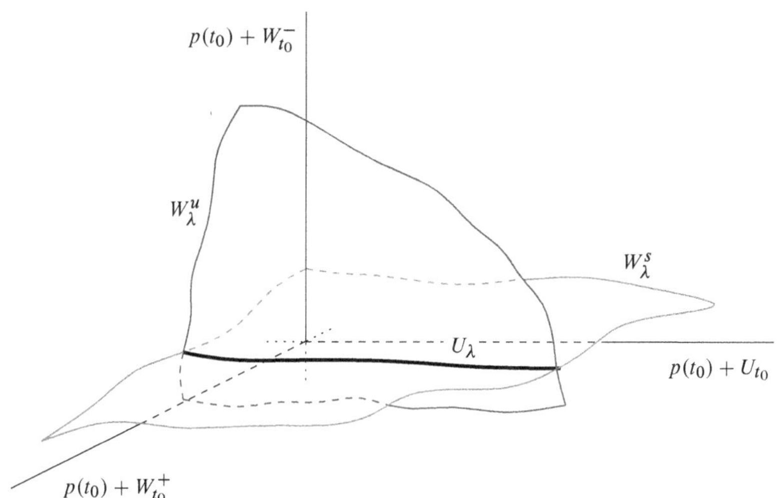

Figure B.4: *Degenerate (homo)heteroclinic orbit.*

R. Abraham and S. Smale (1970). "Nongenericity of Ω-stability." In: *Global Analysis (Proc. Sympos. Pure Math., Vol. XIV, Berkeley, Calif., 1968)*. Amer. Math. Soc., Providence, R.I., pp. 5–8. MR: 0271986. Zbl: 0215.25102 (cit. on p. vi).

V. S. Afraimovich, S. V. Gonchenko, L. M. Lerman, A. L. Shilnikov, and D. V. Turaev (2014). "Scientific heritage of LP Shilnikov." *Regular and Chaotic Dynamics* 19.4, pp. 435–460. MR: 3240978. Zbl: 1353.37001 (cit. on p. 64).

M. A. D. Aguiar, I. S. Labouriau, and A. A. P. Rodrigues (2010). "Switching near a network of rotating nodes." *Dynamical Systems. An International Journal* 25.1, pp. 75–95. MR: 2765449. Zbl: 1186.37060 (cit. on pp. 35, 47).

C. J. Amick and J. F. Toland (1992). "Homoclinic orbits in the dynamic phase-space analogy of an elastic strut." *European J. Appl. Math.* 3.2, pp. 97–114. MR: 1166253. Zbl: 0755.73023 (cit. on p. 141).

D. V. Anosov (1967). "Geodesic flows on closed Riemannian manifolds of negative curvature." *Trudy Mat. Inst. Steklov.* 90, p. 209. MR: 0224110 (cit. on p. iii).

V. Araujo, M. J. Pacifico, E. R. Pujals, and M. Viana (2009). "Singular-hyperbolic attractors are chaotic." *Transactions of the American Mathematical Society* 361.5, pp. 2431–2485. MR: 2471925. Zbl: 1214.37010 (cit. on p. xiv).

A. Arncodo, P. Coullet, and C. Tresser (1981). "Possible new strange attractors with spiral structure." *Communications in Mathematical Physics* 79.4, pp. 573–579. MR: 0623967. Zbl: 0485.58013 (cit. on pp. xiii, 33).

V. I. Arnold, V. S. Afrajmovich, Y. Il'yashenko, and L. P. Shilnikov (1999). *Bifurcation theory and catastrophe theory*. Springer Berlin. MR: 1733750 (cit. on p. 9).

I. Baldomá, O. Castejón, and T. M. Seara (2013). "Exponentially small heteroclinic breakdown in the generic Hopf-Zero singularity." *Journal of Dynamics and Differential Equations* 25.2, pp. 335–392. MR: 3054640. Zbl: 1269.37025 (cit. on pp. 151, 152).

— (2018a). "Breakdown of a 2D heteroclinic connection in the Hopf-Zero singularity (I)." *J. Nonlinear Science* 28.5, pp. 1551–1627. MR: 3846870. Zbl: 1404.34054 (cit. on pp. 151, 152).

— (2018b). "Breakdown of a 2D heteroclinic connection in the Hopf-Zero singularity (II). The generic case." *J. Nonlinear Science* 28.4, pp. 1489–1549. MR: 3817789. Zbl: 06928649 (cit. on pp. 151, 152).

I. Baldomá, S. Ibáñez, and T. M. Seara (Mar. 2019). "Hopf-Zero singularities truly unfold chaos." arXiv: 1903.09023 (cit. on p. 152).

A. Banyaga, R. de la Llave, and C. E. Wayne (1996). "Cohomology equations near hyperbolic points and geometric versions of Sternberg linearization theorem." *The Journal of Geometric Analysis* 6.4, pp. 613–649. MR: 1601409. Zbl: 0918.53016 (cit. on p. 76).

P. G. Barrientos, S. Ibáñez, and J. A. Rodríguez (2016). "Robust cycles unfolding from conservative bifocal homoclinic orbits." *Dynamical Systems. An International Journal* 31.4, pp. 546–579. MR: 3574314. Zbl: 1380.37044 (cit. on pp. 56, 84, 85).

P. G. Barrientos, S. Ibáñez, and J. A. Rodríguez (2011). "Heteroclinic cycles arising in generic unfoldings of nilpotent singularities." *Journal of Dynamics and Differential Equations* 23.4, pp. 999–1028. MR: 2859948. Zbl: 1242.34079 (cit. on pp. xii, xiv, 40, 46, 110, 136, 142).

P. G. Barrientos, Y. Ki, and A. Raibekas (2014). "Symbolic blender-horseshoes and applications." *Nonlinearity* 27.12, p. 2805. MR: 3291132. Zbl: 1336.37025 (cit. on p. 85).

P. G. Barrientos and A. Raibekas (Nov. 2017). "Robust tangencies of large codimension." *Nonlinearity* 30.12, pp. 4369–4409. MR: 3734140. Zbl: 1385.37027 (cit. on p. 70).

P. G. Barrientos, A. Raibekas, and A. A. P. Rodrigues (2019). "Chaos near a reversible homoclinic bifocus." *Dynamical Systems. An International Journal* 34 (3), pp. 504–516 (cit. on pp. 56, 88, 91–93).

G. R. Belitskii (1973). "Functional equations and conjugacy of local diffeomorphisms of a finite smoothness class." *Functional Analysis and Its Applications* 7.4, pp. 268–277. MR: 0331437 (cit. on p. 4).

L. A. Belyakov (1984a). "Bifurcation of systems with homoclinic curve of a saddle-focus with saddle quantity zero." *Mathematical Notes* 36.5, pp. 838–843. MR: 0773804. Zbl: 0598.34035 (cit. on p. 57).

— (1984b). "Bifurcations of systems with a homoclinic curve of the saddle-focus with a zero saddle value." *Mat. Zametki* 36.5, pp. 681–689, 798. MR: 0773804. Zbl: 0561.34019 (cit. on pp. 84, 142).

L. A. Belyakov and L. P. Shilnikov (1990). "Homoclinic curves and complex solitary waves." *Selecta Math. Soviet.* 9.3, pp. 219–228. MR: 1074385. Zbl: 0739.35086 (cit. on p. 142).

M. Benedicks and L. Carleson (1985). "On iterations of $1 - ax^2$ on $(-1, 1)$." *Annals of Mathematics*, pp. 1–25. MR: 0799250. Zbl: 0597.58016 (cit. on pp. viii, 24, 73).

— (1991). "The dynamics of the Hénon map." *Annals of Mathematics*, pp. 73–169. MR: 1087346. Zbl: 0724.58042 (cit. on pp. vi, xi, 24, 29, 30, 73).

P. Berger (2016). "Generic family with robustly infinitely many sinks." *Inventiones mathematicae* 205.1, pp. 121–172. MR: 3514960. Zbl: 1365.37024 (cit. on pp. vii, 31).

— (2017). "Emergence and non-typicality of the finiteness of the attractors in many topologies." *Proceedings of the Steklov Institute of Mathematics* 297.1, pp. 1–27. Zbl: 1381.37035 (cit. on pp. vii, 31).

M. Bessa, M. Carvalho, and A. A. P. Rodrigues (2017). "Quasi-stochastic Bykov attractors." arXiv: 1712.01885 (cit. on p. 48).

M. Bessa, J. Rocha, and P. Varandas (2018). "Uniform hyperbolicity revisited: index of periodic points and equidimensional cycles." *Dynamical Systems. An International Journal* 33.4, pp. 691–707. MR: 3869855. Zbl: 1402.37038 (cit. on p. 74).

M. Bessa and A. A. P. Rodrigues (2016). "Dynamics of conservative Bykov cycles: Tangencies, generalized Cocoon bifurcations and elliptic solutions." *Journal of Differential Equations* 261.2, pp. 1176–1202. MR: 3494394. Zbl: 1360.37130 (cit. on p. 50).

G. D. Birkhoff (1927). *Dynamical systems.* reprinted with an addendum by Jurgen Moser, 1966. Providence, R.I.: American Mathematical Society, pp. xii+305. MR: 0209095 (cit. on pp. iii, x, 7).

R. I. Bogdanov (1975). "Versal deformation of a singular point of a vector field on the plane in the case of zero eigenvalues." *Funkcional Anal. i Priložen.* 9.2, p. 63. MR: 0380878. Zbl: 0447.58009 (cit. on p. 112).

C. Bonatti and L. J. Díaz (2008). "Robust heterodimensional cycles and C^1-generic dynamics." *J. Inst. Math. Jussieu* 7.3, pp. 469–525. MR: 2427422. Zbl: 1156.37004 (cit. on pp. 85, 94).

— (2012). "Abundance of C^1-homoclinic tangencies." *Transactions of the American Mathematical Society* 264, pp. 5111–5148. MR: 2931324. Zbl: 1300.37013 (cit. on pp. 85, 93).

C. Bonatti, L. J. Díaz, and M. Viana (2005). *Dynamics beyond uniform hyperbolicity.* Vol. 102. Encyclopaedia of Mathematical Sciences. Berlin: Springer-

Verlag, pp. xviii+384. MR: 2105774. Zbl: 1060 . 37020 (cit. on pp. xiv, 84).

C. Bonatti, A. Pumariño, and M. Viana (1997). "Lorenz attractors with arbitrary expanding dimension." *Comptes rendus de l'Académie des sciences. Série 1, Mathématique* 325.8, pp. 883–888. MR: 1485910. Zbl: 0896 . 58043 (cit. on p. xiv).

C. Bonatti, L. J. Díaz, and E. R. Pujals (2003). "A C^1-generic dichotomy for diffeomorphisms: weak forms of hyperbolicity or infinitely many sinks or sources." *Annals of Mathematics*, pp. 355–418. MR: 2018925. Zbl: 1049 . 37011 (cit. on p. 74).

P. Bonckaert (1997). "Conjugacy of vector fields respecting additional properties." *Journal of dynamical and control systems* 3.3, pp. 419–432. MR: 1472359. Zbl: 1157 . 37322 (cit. on p. 5).

P. Bonckaert and E. Fontich (2003). "Invariant manifolds of maps close to a product of rotations: close to the rotation axis." *Journal of Differential Equations* 191.2, pp. 490–517. MR: 1978387. Zbl: 1070 . 37016 (cit. on p. 151).

— (2005). "Invariant manifolds of dynamical systems close to a rotation: transverse to the rotation axis." *Journal of Differential Equations* 214.1, pp. 128–155. MR: 2143514. Zbl: 1082 . 34037 (cit. on p. 151).

H. W. Broer (1981). "Formal normal form theorems for vector fields and some consequences for bifurcations in the volume preserving case." In: *Dynamical systems and turbulence, Warwick 1980 (Coventry, 1979/1980)*. Vol. 898. Lecture Notes in Math. Springer, Berlin-New York, pp. 54–74. MR: 654883 (cit. on p. 102).

H. W. Broer, F. Dumortier, S. J. van Strien, and F. Takens (1991). *Structures in dynamics*. Vol. 2. Studies in Mathematical Physics. Amsterdam: North-Holland Publishing Co., pp. xii+309. MR: 1134128. Zbl: 0746 . 58002 (cit. on pp. 101–103, 107, 109, 112, 113).

H. W. Broer, R. Roussarie, and C. Simó (1996). "Invariant circles in the Bogdanov-Takens bifurcation for diffeomorphisms." *Ergodic Theory and Dynamical Systems* 16.6, pp. 1147–1172. MR: 1424392. Zbl: 0876 . 58032 (cit. on p. 71).

H. W. Broer and G. Vegter (1984). "Subordinate 'Sil'nikov bifurcations near some singularities of vector fields having low codimension." *Ergodic Theory and Dynamical Systems* 4.4, pp. 509–525. MR: 0779710. Zbl: 0553 . 58024 (cit. on pp. 147, 149, 151).

I. U. Bronstein and A. Y. Kopanskii (1996). "Normal forms of vector fields satisfying certain geometric conditions." In: *Nonlinear Dynamical Systems and Chaos*. Springer, pp. 79–101. MR: 1391492. Zbl: 0843 . 58110 (cit. on p. 76).

B. Buffoni, A. R. Champneys, and J. F. Toland (1996). "Bifurcation and coalescence of a plethora of homoclinic orbits for a Hamiltonian system." *Journal of Dynamics and Differential Equations* 8.2, pp. 221–279. MR: 1388874. Zbl: 0854.34047 (cit. on p. 141).

V. V. Bykov (1993). "The bifurcations of separatrix contours and chaos." *Physica D* 62.1-4, pp. 290–299. MR: 1207427. Zbl: 0799.58054 (cit. on p. 39).

— (1999). "On systems with separatrix contour containing two saddle-foci." *J. Math. Sci. (New York)* 95.5, pp. 2513–2522. MR: 1712740. Zbl: 1160.37377 (cit. on pp. 39, 51–53).

— (2000). "Orbit structure in a neighborhood of a separatrix cycle containing two saddle-foci." In: *Methods of qualitative theory of differential equations and related topics*. Vol. 200. Amer. Math. Soc. Transl. Ser. 2. Providence, RI: Amer. Math. Soc., pp. 87–97. MR: 1769565. Zbl: 1026.37043 (cit. on pp. 39, 46, 51–53).

J. Carr (1982). *Applications of Centre Manifold Theory*. Vol. 35. Applied Mathematical Sciences. Springer-Verlag New York (cit. on p. 100).

J. C. Ceballos and R. Labarca (1992). "A note on modulus of stability for cycles of the complex type." *Physica D* 55.1-2, pp. 37–44. MR: 1152002. Zbl: 0753.58026 (cit. on p. 9).

A. R. Champneys and J. F. Toland (1993). "Bifurcation of a plethora of multimodal homoclinic orbits for autonomous Hamiltonian systems." *Nonlinearity* 6.5, pp. 665–721. MR: 1240679. Zbl: 0789.58035 (cit. on p. 141).

K.-T. Chen (1963). "Equivalence and decomposition of vector fields about an elementary critical point." *American Journal of Mathematics* 85.4, pp. 693–722. MR: 0160010. Zbl: 0119.07505 (cit. on p. 5).

C. Chicone (2006). *Ordinary differential equations with applications*. Second. Vol. 34. Texts in Applied Mathematics. New York: Springer, pp. xx+636. MR: 2224508. Zbl: 1120.34001 (cit. on p. 163).

S.-N. Chow and J. K. Hale (1982). *Methods of bifurcation theory*. Vol. 251. Grundlehren der Mathematischen Wissenschaften. New York: Springer-Verlag, pp. xv+515. MR: 0660633 (cit. on p. 104).

S.-N. Chow, J. K. Hale, and J. Mallet-Paret (1980). "An example of bifurcation to homoclinic orbits." *Journal of Differential Equations* 37.3, pp. 351–373. MR: 0589997. Zbl: 0439.34035 (cit. on pp. ix, 176).

E. Colli (1998). "Infinitely many coexisting strange attractors." *Annales de l'Institut Henri Poincare-Nonlinear Analysis* 15.5, pp. 539–580. MR: 1643393. Zbl: 0932.37015 (cit. on pp. vii, xi, 30, 31).

W. A. Coppel (1978). *Dichotomies in stability theory*. Vol. 629. Lecture Notes in Mathematics. Berlin: Springer-Verlag, pp. ii+98. MR: 0481196. Zbl: 0376.34001 (cit. on p. 156).

F. Costal and J. A. Rodríguez (1985). "A bifurcation problem to homoclinic orbits for nonautonomous systems." *Journal of Mathematical Analysis and Applications* 105.2, pp. 395–404. MR: 778474. Zbl: 0565.34050 (cit. on p. ix).

— (1988). "Bifurcation to homoclinic orbits and to periodic solutions for nonautonomous three-dimensional systems." *Archive for Rational Mechanics and Analysis* 104.2, pp. 185–194. MR: 959401. Zbl: 0683.34019 (cit. on p. ix).

S. Crovisier and E. R. Pujals (2015). "Essential hyperbolicity and homoclinic bifurcations: a dichotomy phenomenon/mechanism for diffeomorphisms." English. *Inventiones mathematicae* 201.2, pp. 385–517. MR: 3370619. Zbl: 1377.37031 (cit. on p. vii).

B. Deng (1989a). "Exponential expansion with Sil'nikov's saddle-focus." *Journal of Differential Equations* 82.1, pp. 156–173. MR: 1023305. Zbl: 0703.34041 (cit. on p. 64).

— (1989b). "The Sil'nikov problem, exponential expansion, strong λ-lemma, C^1-linearization, and homoclinic bifurcation." *Journal of Differential Equations* 79.2, pp. 189–231. MR: 1000687. Zbl: 0674.34040 (cit. on p. 5).

— (1993). "On Silnikov's Homoclinic-Saddle-Focus Theorem." *Journal of Differential Equations* 102.2, pp. 305–329. MR: 1216731. Zbl: 0780.34029 (cit. on p. 64).

R. L. Devaney (1976a). "Homoclinic orbits in Hamiltonian systems." *Journal of Differential Equations* 21.2, pp. 431–438. MR: 0442990. Zbl: 0343.58005 (cit. on pp. 55, 57, 87, 142).

R. L. Devaney (1976b). "Reversible diffeomorphisms and flows." *Transactions of the American Mathematical Society* 218, pp. 89–113. MR: 0402815. Zbl: 0363.58003 (cit. on pp. xiv, 55).

— (1977). "Blue sky catastrophes in reversible and Hamiltonian systems." *Indiana University Mathematics Journal* 26.2, pp. 247–263. MR: 0431274. Zbl: 0362.58006 (cit. on pp. 56, 57, 87).

F. Drubi (2009). "Synchronization and chaos in coupled systems: The model of two coupled Brusselators." Phd. thesis. Universidad de Oviedo (cit. on p. 138).

F. Drubi, S. Ibáñez, and D. Rivela (2019a). "A formal classification of Hopf-Bogdanov-Takens singularities of codimension three" (cit. on pp. 139, 153).

— (2019b). "Chaotic behaviour in the unfolding of Hopf-Bogdanov-Takens singularities" (cit. on p. 153).

F. Drubi, S. Ibáñez, and J. A. Rodríguez (2007). "Coupling leads to chaos." *Journal of Differential Equations* 239.2, pp. 371–385. MR: 2344277. Zbl: 1133.34027 (cit. on pp. v, 108, 147).

E. Dufraine (2001). "Some topological invariants for three-dimensional flows." *Chaos: An Interdisciplinary Journal of Nonlinear Science* 11.3, pp. 443–448. MR: 1855419. Zbl: 0971.37008 (cit. on p. 10).

F. Dumortier and S. Ibáñez (1996). "Nilpotent singularities in generic 4-parameter families of 3-dimensional vector fields." *Journal of Differential Equations* 127.2, pp. 590–647. MR: 1389412. Zbl: 0865.58032 (cit. on pp. xii, 99, 108, 109, 123).

— (1998). "Singularities of vector fields on $\mathbf{R}^3$." *Nonlinearity* 11.4, pp. 1037–1047. MR: 1632586. Zbl: 0907.58051 (cit. on pp. 99, 100, 148).

— (1999). "Codimension 4 singularities of reflectionally symmetric planar vector fields." *Publ. Mat.* 43.2, pp. 501–533. MR: 1744620. Zbl: 0964.58034 (cit. on p. 99).

F. Dumortier, S. Ibáñez, and H. Kokubu (2001a). "New aspects in the unfolding of the nilpotent singularity of codimension three." *Dynamical Systems. An International Journal* 16.1, pp. 63–95. MR: 1835907. Zbl: 0993.37026 (cit. on p. 40).

— (2001b). "New aspects in the unfolding of the nilpotent singularity of codimension three." *Dynamical Systems. An International Journal* 16.1, pp. 63–95. MR: 1835907. Zbl: 0993.37026 (cit. on pp. 119, 122–124).

— (2006). "Cocoon bifurcation in three-dimensional reversible vector fields." *Nonlinearity* 19.2, pp. 305–328. MR: 2199390. Zbl: 1107.34039 (cit. on p. 123).

F. Dumortier, J. Llibre, and J. C. Artés (2006). *Qualitative Theory of Planar Differential Systems*. Universitext. Springer Berlin-Heidelberg. MR: 2256001 (cit. on p. 103).

F. Dumortier (1977). "Singularities of vector fields on the plane." *Journal of Differential Equations* 23.1, pp. 53–106. MR: 0650816. Zbl: 0346.58002 (cit. on p. 99).

F. Dumortier, S. Ibánez, and H. Kokubu (2005). "Cocoon bifurcation in three-dimensional reversible vector fields." *Nonlinearity* 19.2, p. 305. MR: 2199390. Zbl: 1107.34039 (cit. on p. 46).

F. Dumortier, S. Ibánez, H. Kokubu, and C. Simó (2013). "About the unfolding of a Hopf-zero singularity." *Discrete Contin. Dyn. Syst* 33.10, pp. 4435–4471. MR: 3049086. Zbl: 1283.34039 (cit. on pp. 40, 46, 151, 152).

F. Dumortier, H. Kokubu, and H. Oka (1995). "A degenerate singularity generating geometric Lorenz attractors." *Ergodic Theory and Dynamical Systems* 15.5, pp. 833–856. MR: 1356617. Zbl: 0836.58030 (cit. on p. xiii).

F. Fernández-Sánchez, E. Freire, L. Pizarro, and A. J. Rodríguez-Luis (2003). "A model for the analysis of the dynamical consequences of a nontransversal intersection of the two-dimensional manifolds involved in a T-point." *Physics Letters A* 320.2-3, pp. 169–179. Zbl: 1065.37019 (cit. on p. 39).

F. Fernández-Sánchez, E. Freire, and A. J. Rodríguez-Luis (2002). "T-points in a Z2-symmetric electronic oscillator. (I) Analysis." *Nonlinear Dynamics* 28.1, pp. 53–69. MR: 1900510 (cit. on pp. 39, 40, 46).

A. C. Fowler and C. T. Sparrow (1991). "Bifocal homoclinic orbits in four dimensions." *Nonlinearity* 4.4, p. 1159. MR: 1138131. Zbl: 0741.34017 (cit. on pp. xiv, 55, 61, 64).

P. Gaspard (1993). "Local birth of homoclinic chaos." *Physica D: Nonlinear Phenomena* 62.1-4, pp. 94–122. MR: 1207421. Zbl: 0786.34051 (cit. on p. 151).

P. Gaspard, R. Kapral, and G. Nicolis (1984). "Bifurcation phenomena near homoclinic systems: a two-parameter analysis." *Journal of Statistical Physics* 35.5-6, pp. 697–727. MR: 0749843. Zbl: 0588.58055 (cit. on p. 7).

P. Glendinning and C. T. Sparrow (1984). "Local and global behavior near homoclinic orbits." *Journal of Statistical Physics* 35.5-6, pp. 645–696. MR: 0749842. Zbl: 0588.58041 (cit. on pp. 7, 20, 22, 23).

— (1986). "T-points: a codimension two heteroclinic bifurcation." *Journal of Statistical Physics* 43.3-4, pp. 479–488. MR: 0845725. Zbl: 0635.58031 (cit. on pp. 39, 40).

M. Golubitsky, I. Stewart, and D. G. Schaeffer (2012). *Singularities and groups in bifurcation theory*. Vol. 2. Springer Science & Business Media. MR: 0771477. Zbl: 0691.58003 (cit. on p. 33).

S. V. Gonchenko, V. S. Gonchenko, and J. C. Tatjer (2007). "Bifurcations of three-dimensional diffeomorphisms with non-simple quadratic homoclinic tangencies and generalized Hénon maps." *Regular and Chaotic Dynamics* 12.3, pp. 233–266. MR: 2350324. Zbl: 1229.37040 (cit. on pp. vii, 65, 68, 70).

S. V. Gonchenko, D. V. Turaev, P. Gaspard, and G. Nicolis (1997). "Complexity in the bifurcation structure of homoclinic loops to a saddle-focus." *Nonlinearity* 10.2, p. 409. MR: 1438259. Zbl: 0905.34042 (cit. on p. 7).

S. V. Gonchenko, D. V. Turaev, and L. P. Shilnikov (1993). "On the existence of Newhouse regions in a neighborhood of systems with a structurally unstable homoclinic Poincaré curve (the multidimensional case)." *Dokl. Akad. Nauk* 329.4, pp. 404–407. MR: 1238829(94i:58147) (cit. on p. 68).

J. Guckenheimer (1981). "On a codimension two bifurcation." In: *Dynamical systems and turbulence, Warwick 1980 (Coventry, 1979/1980)*. Vol. 898. Lecture Notes in Math. Springer, Berlin-New York, pp. 99–142. MR: 654886. Zbl: 0482.58006 (cit. on p. 151).

J. Guckenheimer and P. J. Holmes (2002). *Nonlinear oscillations, dynamical systems, and bifurcations of vector fields*. 7th. Vol. 42. Applied Mathematical Sciences. New York: Springer-Verlag, pp. xvi+459. MR: 0709768. Zbl: 0515.34001 (cit. on pp. 102, 119, 138, 147, 148, 151, 152).

J. Guckenheimer and R. F. Williams (1979). "Structural stability of Lorenz attractors." *Inst. Hautes Études Sci. Publ. Math.* 50, pp. 59–72. MR: 556582. Zbl: 0436.58018 (cit. on p. xiii).

J. Härterich (1998). "Cascades of reversible homoclinic orbits to a saddle-focus equilibrium." *Physica D: Nonlinear Phenomena* 112.1-2, pp. 187–200. MR: 1605837 (cit. on pp. 55, 57, 60, 87, 88).

M. Hénon (1976). "A two-dimensional mapping with a strange attractor." In: *The Theory of Chaotic Attractors*. Springer, pp. 94–102. MR: 0422932 (cit. on pp. v, 24).

M. W. Hirsch, C. C. Pugh, and M. Shub (1977). *Invariant manifolds*. Lecture Notes in Mathematics, Vol. 583. Berlin: Springer-Verlag, pp. ii+149. MR: 0501173. Zbl: 0355.58009 (cit. on p. 69).

H. Hofer and J. F. Toland (1984). "Homoclinic, heteroclinic, and periodic orbits for a class of indefinite Hamiltonian systems." *Math. Ann.* 268.3, pp. 387–403. MR: 0751737. Zbl: 0569.70017 (cit. on p. 141).

P. J. Holmes (1980). "A strange family of three-dimensional vector fields near a degenerate singularity." *Journal of Differential Equations* 37.3, pp. 382–403. MR: 0589999. Zbl: 0421.58016 (cit. on p. 35).

A. J. Homburg (2002). "Periodic attractors, strange attractors and hyperbolic dynamics near homoclinic orbits to saddle-focus equilibria." *Nonlinearity* 15.4, p. 1029. MR: 1912285. Zbl: 1017.37012 (cit. on pp. xi, 7, 8, 31, 33).

A. J. Homburg and J. Knobloch (2010). "Switching homoclinic networks." *Dynamical Systems. An International Journal* 25.3, pp. 351–358. MR: 2731619. Zbl: 1204.37028 (cit. on p. 35).

A. J. Homburg and J. S. W. Lamb (2006). "Symmetric homoclinic tangles in reversible systems." *Ergodic theory and dynamical systems* 26.6, pp. 1769–1789. MR: 2279265. Zbl: 1132.37013 (cit. on p. 93).

A. J. Homburg and B. Sandstede (2010). "Homoclinic and heteroclinic bifurcations in vector fields." *Handbook of dynamical systems* 3, pp. 379–524. Zbl: 1243.37024 (cit. on p. 31).

B. R. Hunt and V. Y. Kaloshin (2010). "Prevalence." In: *Handbook of dynamical systems*. Vol. 3. Elsevier, pp. 43–87. Zbl: 1317.37013 (cit. on p. 25).

S. Ibáñez and A. A. P. Rodrigues (2015). "On the dynamics near a homoclinic network to a bifocus: switching and horseshoes." *International Journal of Bifurcation and Chaos* 25.11, p. 1530030. MR: 3416211. Zbl: 1327.34078 (cit. on pp. xiv, 35, 55, 60, 64, 88).

S. Ibáñez and J. A. Rodríguez (1995). "Sil'nikov bifurcations in generic 4-unfoldings of a codimension-4 singularity." *Journal of Differential Equations* 120.2, pp. 411–428. MR: 1347350 (cit. on p. xi).

— (2005). "Shilnikov configurations in any generic unfolding of the nilpotent singularity of codimension three on $\mathbb{R}^3$." *Journal of Differential Equations* 208.1, pp. 147–175. MR: 2107297 (cit. on pp. xii, 40, 46, 128).

M. V. Jakobson (1981). "Absolutely continuous invariant measures for one-parameter families of one-dimensional maps." *Communications in Mathematical Physics* 81.1, pp. 39–88. MR: 0630331. Zbl: 0497.58017 (cit. on pp. viii, 24).

J. Jones, W. C. Troy, and A. D. MacGillivray (1992). "Steady solutions of the Kuramoto-Sivashinsky equation for small wave speed." *Journal of Differential Equations* 96.1, pp. 28–55. MR: 1153308 (cit. on pp. 122, 123).

A. Katok and B. Hasselblatt (1995). *Introduction to the modern theory of dynamical systems*. Vol. 54. Cambridge university press. MR: 1326374. Zbl: 0878.58020 (cit. on p. viii).

U. Kirchgraber and K. J. Palmer (1990). *Geometry in the Neighborhood of Invariant Manifolds of Maps and Flows and Linearization*. Pitman research notes in mathematics series. Longman Scientific & Technical. MR: 1068954. Zbl: 0746.58008 (cit. on p. 101).

S. Kiriki, Y. Nakano, and T. Soma (2017). "Non-trivial wandering domains for heterodimensional cycles." *Nonlinearity* 30.8, p. 3255. MR: 3685668. Zbl: 1382.37049 (cit. on pp. 73, 74).

V. Kirk and M. Silber (1994). "A competition between heteroclinic cycles." *Nonlinearity* 7.6, p. 1605. MR: 1304441. Zbl: 0815.34033 (cit. on p. 35).

J. Knobloch (1997). "Bifurcation of degenerate homoclinics in reversible and conservative systems." *J. Dyn. Diff. Eq.* 9.3, pp. 427–444. MR: 1464081. Zbl: 0884.34057 (cit. on p. 186).

— (2004). "Lin's method for discrete and continuous dynamical systems and applications." thesis. T. U. Ilmenau (cit. on pp. 178, 186, 188).

J. Knobloch, J. S. W. Lamb, and K. N. Webster (2005). "Shift dynamics in a neighborhood of a T-point containing two saddle-foci" (cit. on p. 48).

J. Knobloch, J. S. W. Lamb, and K. N. Webster (2014). "Using Lin's method to solve Bykov's problems." *Journal of Differential Equations* 257.8, pp. 2984–3047. MR: 3249278. Zbl: 1312.34083 (cit. on pp. 46, 53).

H. Kokubu (1991). "Heteroclinic bifurcations associated with different saddle indices." In: *Dynamical systems and related topics*. Vol. 9. Adv. Ser. Dynam. Systems. World Sci. Publ., River Edge, NJ, pp. 236–260. MR: 1164893 (cit. on p. 39).

H. Kokubu, D. Wilczak, and P. Zgliczyński (2007). "Rigorous verification of cocoon bifurcations in the Michelson system." *Nonlinearity* 20.9, pp. 2147–2174. MR: 2351028. Zbl: 1126.37035 (cit. on p. 123).

O. Koltsova and L. M. Lerman (2009). "Hamiltonian dynamics near nontransverse homoclinic orbit to saddle-focus equilibrium." *Discrete and Contin-*

uous Dynamical Systems Series A 25.3, p. 883. MR: 2533981. Zbl: 1188. 37029 (cit. on p. 55).

Y. Kuramoto and T. Tsuzuki (Feb. 1976). "Persistent Propagation of Concentration Waves in Dissipative Media Far from Thermal Equilibrium." *Progress of Theoretical Physics* 55.2, pp. 356–369 (cit. on pp. 121, 122).

Y. A. Kuznetsov (2004). *Elements of applied bifurcation theory*. Third. Vol. 112. Applied Mathematical Sciences. New York: Springer-Verlag, pp. xxii+631. MR: 2071006. Zbl: 1082.37002 (cit. on pp. 102, 107, 119, 138, 147, 152).

Y. A. Kuznetsov and H. G. E. Meijer (2019). "Center Manifold Reduction for Local Bifurcations." In: *Numerical Bifurcation Analysis of Maps: From Theory to Software*. Cambridge Monographs on Applied and Computational Mathematics. Cambridge University Press, 185?216 (cit. on p. 103).

I. S. Labouriau and A. A. P. Rodrigues (2012). "Global generic dynamics close to symmetry." *Journal of Differential Equations* 253.8, pp. 2527–2557. MR: 2950462. Zbl: 1255.34045 (cit. on pp. 35, 46).

— (2015). "Dense heteroclinic tangencies near a Bykov cycle." *Journal of Differential Equations* 259.11, pp. 5875–5902. MR: 3397312. Zbl: 1326. 34078 (cit. on pp. 40, 48–51).

— (2016). "Global bifurcations close to symmetry." *Journal of Mathematical Analysis and Applications* 444.1, pp. 648–671. MR: 3523396. Zbl: 1364. 37115 (cit. on pp. 46, 48).

— (2017). "On Takens Last Problem: tangencies and time averages near heteroclinic networks." *Nonlinearity* 30.5, p. 1876. MR: 3639293. Zbl: 1370. 34073 (cit. on p. 51).

J. S. W. Lamb, M.-A. Teixeira, and K. N. Webster (2005). "Heteroclinic bifurcations near Hopf-zero bifurcation in reversible vector fields in $\mathbb{R}^3$." *Journal of Differential Equations* 219.1, pp. 78–115. MR: 2181031. Zbl: 1090.34033 (cit. on pp. 40, 46–48).

Y.-T. Lau (1992). "The "Cocoon" bifurcations in three-dimensonal systems with two fixed points." *International Journal of Bifurcation and Chaos* 02.03, pp. 543–558. MR: 1192697. Zbl: 0874.58059 (cit. on pp. 122, 123).

L. M. Lerman (1991). "Complex dynamics and bifurcations in a Hamiltonian system having a transversal homoclinic orbit to a saddle focus." *Chaos: An Interdisciplinary Journal of Nonlinear Science* 1.2, pp. 174–180. MR: 1135905. Zbl: 0899.58053 (cit. on pp. 57, 79).

— (1997). "Homo-and heteroclinic orbits, hyperbolic subsets in a one-parameter unfolding of a Hamiltonian system with heteroclinic contour with two saddle-foci." *Regular and Chaotic Dynamics* 2.3-4, pp. 139–155. MR: 1702345. Zbl: 0945.37017 (cit. on p. 79).

L. M. Lerman (2000). "Dynamical phenomena near a saddle-focus homoclinic connection in a Hamiltonian system." *Journal of Statistical Physics* 101.1-2, pp. 357–372. MR: 1807550. Zbl: 1003.37036 (cit. on pp. 79, 83, 84).

X.-B. Lin (1990). "Using Mel'nikov's method to solve Šilnikov's problems." *Proc. Roy. Soc. Edinburgh Sect. A* 116.3-4, pp. 295–325. MR: 1084736 (cit. on p. 176).

Z. Lin and Y.-X. Lin (2000). *Linear systems exponential dichotomy and structure of sets of hyperbolic points*. World Scientific. MR: 1763235. Zbl: 0953.34003 (cit. on p. 156).

E. N. Lorenz (1963). "Deterministic nonperiodic flow." *Journal of the atmospheric sciences* 20.2, pp. 130–141. Zbl: 06851676 (cit. on pp. v, ix, 3).

V. V. Lyčagin (1977). "On sufficient orbits of a group of contact diffeomorphisms." *Mathematics of the USSR-Sbornik* 33.2, p. 223 (cit. on p. 76).

N. G. Markley (2011). *Principles of differential equations*. Vol. 67. John Wiley & Sons. MR: 2063923. Zbl: 1057.34001 (cit. on p. 155).

J. L. Massera and J. J. Schäffer (1966). *Linear differential equations and function spaces*. Pure and Applied Mathematics, Vol. 21. New York: Academic Press, pp. xx+404. MR: 0212324 (cit. on p. 156).

V. K. Melnikov (1963). "On the stability of the center for time-periodic perturbations." *Trans. Moscow Math. Soc* 12.1, pp. 1–57 (cit. on p. ix).

D. Michelson (1986). "Steady solutions of the Kuramoto-Sivashinsky equation." *Physica D: Nonlinear Phenomena* 19.1, pp. 89–111. MR: 0840029. Zbl: 0603.35080 (cit. on pp. 121, 122).

L. Mora and M. Viana (1993). "Abundance of strange attractors." *Acta Mathematica* 171.1, pp. 1–71. MR: 1237897. Zbl: 0815.58016 (cit. on pp. vi, vii, ix–xi, 25, 29, 30, 71, 73).

C. A. Morales, M. J. Pacifico, and E. R. Pujals (1999). "Singular hyperbolic systems." *Proceedings of the American Mathematical Society* 127.11, pp. 3393–3401. MR: 1610761. Zbl: 0924.58068 (cit. on p. xiv).

— (2004). "Robust transitive singular sets for 3-flows are partially hyperbolic attractors or repellers." *Annals of mathematics*, pp. 375–432. MR: 2123928. Zbl: 1071.37022 (cit. on p. xiv).

J. Moser (1958). "On the generalization of a theorem of A. Liapounoff." *Communications on Pure and Applied Mathematics* 11.2, pp. 257–271. MR: 0096021 (cit. on p. 76).

S. E. Newhouse (1970). "Nondensity of axiom A(a) on S^2." In: *Global Analysis (Proc. Sympos. Pure Math., Vol. XIV, Berkeley, Calif., 1968)*. Amer. Math. Soc., Providence, R.I., pp. 191–202. MR: 0277005 (cit. on p. vi).

— (1974). "Diffeomorphisms with infinitely many sinks." *Topology* 13, pp. 9–18. MR: 0339291. Zbl: 0275.58016 (cit. on p. vi).

— (1979). "The abundance of wild hyperbolic sets and non-smooth stable sets for diffeomorphisms." *Publications Mathématiques de l'IHÉS* 50, pp. 101–151. MR: 0556584. Zbl: 0445.58022 (cit. on pp. vi, 26).

— (2017). "On a differentiable linearization theorem of Philip Hartman. Modern Theory of Dynamical Systems: A Tribute to Dmitry Victorovich Anosov." *Contemporary Mathematics* 692, pp. 209–262. MR: 3666075 (cit. on pp. 4, 11).

B. E. Oldeman, B. Krauskopf, and A. R. Champneys (2001). "Numerical unfoldings of codimension-three resonant homoclinic flip bifurcations." *Nonlinearity* 14.3, p. 597. MR: 1830909. Zbl: 0990.37039 (cit. on p. 40).

I. M. Ovsyannikov and L. P. Shilnikov (1987). "On systems with a saddle-focus homoclinic curve." *Sbornik: Mathematics* 58.2, pp. 557–574. MR: 0867343. Zbl: 0628.58044 (cit. on pp. 7, 8, 31).

M. J. Pacifico, A. Rovella, and M. Viana (1998). "Infinite-modal maps with global chaotic behavior." *Annals of mathematics*, pp. 441–484. MR: 1668551. Zbl: 0916.58029 (cit. on pp. xiii, 33).

J. Palis (1978). "A differentiable invariant of topological conjugacies and moduli of stability." *Asterisque* 51, pp. 335–346. MR: 0494283. Zbl: 0396.58015 (cit. on pp. 9, 10).

J. Palis and M. Viana (1994). "High dimension diffeomorphisms displaying infinitely many periodic attractors." *Ann. of Math. (2)* 140.1, pp. 207–250. MR: 1289496. Zbl: 0817.58004 (cit. on p. 68).

J. Palis (2000). "A global view of dynamics and a conjecture on the denseness of finitude of attractors." *Astérisque* 261.xiiixiv, pp. 335–347. MR: 1755446. Zbl: 1044.37014 (cit. on pp. vi, 30).

J. Palis and F. Takens (1993). "Hyperbolicity and Sensitive Chaotic Dynamics at Homoclinic Bifurcations." *Cambridge Studies in Advanced Mathematics* 35. MR: 1237641. Zbl: 0790.58014 (cit. on pp. vi, 26, 30).

K. J. Palmer (1984). "Exponential dichotomies and transversal homoclinic points." *Journal of Differential Equations* 55.2, pp. 225–256. MR: 0764125. Zbl: 0508.58035 (cit. on pp. 156, 167, 171).

— (1988). "Exponential dichotomies and Fredholm operators." *Proceedings of the American Mathematical Society* 104.1, pp. 149–156. MR: 0958058. Zbl: 0675.34006 (cit. on p. 167).

— (1996). "Shadowing and Silnikov chaos." *Nonlinear Anal.* 27.9, pp. 1075–1093. MR: 1406280. Zbl: 0858.58034 (cit. on p. 171).

— (2000). *Shadowing in dynamical systems*. Vol. 501. Mathematics and its Applications. Dordrecht: Kluwer Academic Publishers, pp. xiv+299. MR: 1885537. Zbl: 0997.37001 (cit. on pp. 156, 164, 171).

H. Poincaré (1890). "Sur le probleme des trois corps et les équations de la dynamique." *Acta Mathematica* 13, pp. 1–270 (cit. on p. iii).

C. C. Pugh and C. Robinson (1983). "The C^1 closing lemma, including Hamiltonians." *Ergodic Theory and Dynamical Systems* 3.2, pp. 261–313. MR: 0742228. Zbl: 0548.58012 (cit. on p. viii).

E. R. Pujals and M. Sambarino (2000). "Homoclinic tangencies and hyperbolicity for surface diffeomorphisms." *Ann. of Math. (2)* 151.3, pp. 961–1023. MR: 1779562(2001m:37057). Zbl: 0959.37040 (cit. on p. vii).

A. Pumariño, J. A. Rodríguez, J. C. Tatjer, and E. Vigil (2013). "Piecewise linear bidimensional maps as models of return maps for 3D diffeomorphisms." In: *Progress and Challenges in Dynamical Systems*. Springer, pp. 351–366. MR: 3111790. Zbl: 1348.37030 (cit. on pp. xiii, 72).

A. Pumariño and J. A. Rodríguez (1997). *Coexistence and persistence of strange attractors*. Vol. 1658. Lecture Notes in Mathematics. Berlin: Springer-Verlag, pp. viii+195. MR: 1456717 (cit. on pp. x, xi, 7, 8, 28–30).

— (2001). "Coexistence and persistence of infinitely many strange attractors." *Ergodic Theory and Dynamical Systems* 21.5, pp. 1511–1523. MR: 1855845. Zbl: 1073.37514 (cit. on pp. xi, 7, 8, 28, 30).

A. Pumariño, J. A. Rodríguez, J. C. Tatjer, and E. Vigil (2014). "Expanding Baker maps as models for the dynamics emerging from 3D-homoclinic bifurcations." *Discrete and Continuous Dynamical Systems Series B* 19.2. MR: 3170197. Zbl: 1291.37034 (cit. on pp. viii, xiii, 72).

— (2015). "Chaotic dynamics for two-dimensional tent maps." *Nonlinearity* 28.2, p. 407. MR: 3303174. Zbl: 1351.37095 (cit. on pp. viii, xiii, 72).

A. Pumariño, J. A. Rodríguez, and E. Vigil (2017). "Renormalizable expanding baker maps: coexistence of strange attractors." *Discrete and Continuous Dynamical Systems Serie A* 37.3, pp. 1651–1678. MR: 3640568. Zbl: 1379.37077 (cit. on pp. viii, xiii, 72).

— (2018). "Renormalization of two-dimensional piecewise linear maps: Abundance of 2-D strange attractors." *Discrete and Continuous Dynamical Systems Serie A* 38.2, pp. 941–966. MR: 3721882. Zbl: 1377.37041 (cit. on pp. viii, xiii, 72).

— (2019). "Persistent two-dimensional strange attractors for a two-parameter family of Expanding Baker Maps." *Discrete and Continuous Dynamical Systems Series B* 24.2. MR: 3927448 (cit. on pp. viii, xiii, 72).

A. Pumariño and J. C. Tatjer (2006). "Dynamics near homoclinic bifurcations of three-dimensional dissipative diffeomorphisms." *Nonlinearity* 19.12, p. 2833. MR: 2273761. Zbl: 1111.37013 (cit. on pp. viii, xiii, 72).

— (2007). "Attractors for return maps near homoclinic tangencies of three-dimensional dissipative diffeomorphisms." *Discrete and Continuous Dynamical Systems Series B* 8.4, p. 971. MR: 2342131. Zbl: 1134.37022 (cit. on pp. viii, xiii, 72).

C. Robinson (1983). "Bifurcation to infinitely many sinks." *Communications in Mathematical Physics* 90.3, pp. 433–459. MR: 0719300. Zbl: 0531.58035 (cit. on pp. vi, 26).

A. A. P. Rodrigues (2013a). "Persistent switching near a heteroclinic model for the geodynamo problem." *Chaos, Solitons & Fractals* 47, pp. 73–86. MR: 3021826. Zbl: 1258.86008 (cit. on p. 35).

— (2013b). "Repelling dynamics near a Bykov cycle." *Journal of Dynamics and Differential Equations* 25.3, pp. 605–625. MR: 3095268. Zbl: 1279.37023 (cit. on pp. 35, 40, 46, 47).

— (2015). "Moduli for heteroclinic connections involving saddle-foci and periodic solutions." *Disc. Conti. Dynam. Systems A* 35.7, pp. 3155–3182. MR: 3343558. Zbl: 1366.37052 (cit. on pp. 9, 10).

— (2016). "Is there switching without suspended horseshoes?" *Bol. Soc. Port. Mat. No. 74, 61–79*. MR: 3642311 (cit. on pp. 35, 36).

— (2017). "Attractors in complex networks." *Chaos: An Interdisciplinary Journal of Nonlinear Science* 27.10, p. 103105. MR: 3708190. Zbl: 1388.37081 (cit. on p. 21).

— (2018). "Strange attractors and non wandering domains near a homoclinic cycle to a bifocus." arXiv: 1801.03831 (cit. on pp. 56, 64, 67, 74).

A. A. P. Rodrigues and I. S. Labouriau (2014). "Spiralling dynamics near heteroclinic networks." *Physica D: Nonlinear Phenomena* 268, pp. 34–49. MR: 3145352. Zbl: 1286.37033 (cit. on p. 35).

J. A. Rodríguez (1986). "Bifurcation to homoclinic connections of the focus-saddle type." *Arch. Rational Mech. Anal.* 93.1, pp. 81–90. MR: 0822337. Zbl: 0594.34068 (cit. on p. xi).

D. Ruelle and F. Takens (1971). "On the nature of turbulence." *Comm. Math. Phys.* 20, pp. 167–192. MR: 0284067. Zbl: 0223.76041 (cit. on p. v).

V. S. Samovol (1979). "Linearization of systems of differential equations in the neighborhood of invariant toroidal manifolds." *Trudy Moskovskogo Matematicheskogo Obshchestva* 38, pp. 187–219. MR: 0544939 (cit. on p. 4).

— (1982). "Equivalence of systems of differential equations in the neighborhood of a singular point." *Trudy Moskovskogo Matematicheskogo Obshchestva* 44, pp. 213–234. MR: 0656287. Zbl: 0526.34025 (cit. on p. 4).

— (1988). "On smooth linearization of systems of differential equations in a neighbourhood of a saddle singular point." *Russian Mathematical Surveys* 43.4, pp. 231–232. MR: 0969587. Zbl: 0717.34016 (cit. on p. 4).

B. Sandstede (1993). "Verzweigungstheorie homokliner Verdopplungen." PhD thesis, Report No. 7. Berlin: Institut for Angewandte Analysis und Stochastik. Zbl: 0850.58012 (cit. on pp. 1, 176, 178).

G. R. Sell (1985). "Smooth linearization near a fixed point." *American Journal of Mathematics*, pp. 1035–1091. MR: 0805804. Zbl: 0574.34025 (cit. on p. 4).

M. V. Shashkov and D. V. Turaev (1999). "An existence theorem of smooth nonlocal center manifolds for systems close to a system with a homoclinic loop." *Journal of nonlinear science* 9.5, pp. 525–573. MR: 1707981. Zbl: 0940.37010 (cit. on p. 4).

L. P. Shilnikov (1965). "A case of the existence of a denumerable set of periodic motions." *Sov. Math. Dokl.* 6, pp. 163–166 (cit. on pp. 7, 10).

— (1967). "Existence of a denumerable set of periodic motions in a four-dimensional space in an extended neighborhood of a saddle-focus." *Soviet Math. Dokl.* 8.1, pp. 54–58. Zbl: 0155.41805 (cit. on pp. ix, xiv, 7, 55, 57, 64).

— (1970). "A contribution to the problem of the structure of an extended neighborhood of a rough equilibrium state of saddle-focus type." *Math. USSR Sbornik* 10, pp. 91–102 (cit. on pp. x, xiv, 7, 19, 55, 64).

L. P. Shilnikov, A. L. Shilnikov, D. V. Turaev, and L. O. Chua (2001). "Methods Of Qualitative Theory In Nonlinear Dynamics (Part II). World Sci." *Singapore, New Jersey, London, Hong Kong*. MR: 1884710. Zbl: 1046.34003 (cit. on p. 21).

C. P. Simon (1972). "A 3-dimensional Abraham–Smale example." *Proceedings of the American Mathematical Society* 34, pp. 629–630. MR: 0295391. Zbl: 0259.58006 (cit. on p. vi).

S. Smale (1967). "Differentiable dynamical systems." *Bull. Amer. Math. Soc.* 73, pp. 747–817. MR: 0228014. Zbl: 0202.55202 (cit. on pp. iii, v, 7).

S. Sternberg (1957). "Local contractions and a theorem of Poincaré." *American Journal of Mathematics* 79.4, pp. 809–824. MR: 0096853. Zbl: 0080.29902 (cit. on p. 5).

— (1958). "On the structure of local homeomorphisms of Euclidean n-space, II." *American Journal of Mathematics* 80.3, pp. 623–631. MR: 0096854. Zbl: 0083.31406 (cit. on pp. 4, 26, 28).

S. J. van Strien (1979). "Center Manifolds are not C^∞." eng. *Mathematische Zeitschrift* 166, pp. 143–146. MR: 0525618. Zbl: 0403.58021 (cit. on p. 101).

F. Takens (1974a). "Forced oscillations and bifurcations." *Comm. Math. Inst. Rijksuniv. Utrecht* 3, pp. 1–59. MR: 0478235. Zbl: 1156.37315 (cit. on pp. 109, 112).

— (1974b). "Singularities of vector fields." *Inst. Hautes Études Sci. Publ. Math.* 43, pp. 47–100. MR: 0339292. Zbl: 0279.58009 (cit. on pp. 99, 102, 108, 109, 112, 147, 148, 152).

J. C. Tatjer (2001). "Three-dimensional dissipative diffeomorphisms with homoclinic tangencies." *Ergodic theory and Dynamical systems* 21.1, pp. 249–302. MR: 1826668. Zbl: 0972.37013 (cit. on pp. vii, 65, 68–70).

Y. Togawa (1987). "A modulus of 3-dimensional vector fields." *Ergodic Theory and Dynamical Systems* 7.2, pp. 295–301. MR: 0896800. Zbl: 0645.58027 (cit. on pp. 9, 10).

C. Tresser (1984). "About some theorems by L. P. 'Sil'nikov." *Ann. Inst. H. Poincaré Phys. Théor.* 40.4, pp. 441–461. MR: 0757766 (cit. on pp. x, 7, 10, 19, 35).

M. Triestino (2014). "Généricité au sens probabiliste dans les difféomorphismes du cercle." *Ensaios Matemáticos* 27, pp. 1–98. MR: 3310019. Zbl: 1368.37003 (cit. on p. 25).

W. Tucker (2002). "A rigorous ODE solver and Smale's 14th problem." *Found. Comput. Math.* 2.1, pp. 53–117. MR: 1870856. Zbl: 1047.37012 (cit. on p. xiii).

J. B. Van den Berg, S. A. van Gils, and T. P. P. Visser (2003). "Parameter dependence of homoclinic solutions in a single long Josephson junction." *Nonlinearity* 16.2, p. 707. MR: 1959106. Zbl: 1042.34081 (cit. on p. 40).

A. Vanderbauwhede and B. Fiedler (1992). "Homoclinic period blow-up in reversible and conservative systems." *Zeitschrift für angewandte Mathematik und Physik ZAMP* 43.2, pp. 292–318. MR: 1162729. Zbl: 0762.34023 (cit. on p. 87).

M. Viana (1993). "Strange attractors in higher dimensions." *Boletim da Sociedade Brasileira de Matemática-Bulletin/Brazilian Mathematical Society* 24.1, pp. 13–62. MR: 1224299. Zbl: 0784.58044 (cit. on pp. 68, 71).

— (1997). "Multidimensional nonhyperbolic attractors." *Publications Mathématiques de l'Institut des Hautes Études Scientifiques* 85.1, pp. 63–96. MR: 1471866. Zbl: 1037.37016 (cit. on p. 71).

Q. Wang and L.-S. Young (2001). "Strange attractors with one direction of instability." *Communications in Mathematical Physics* 218.1, pp. 1–97. MR: 1824198. Zbl: 0996.37040 (cit. on p. 30).

— (2008). "Toward a theory of rank one attractors." *Annals of Mathematics*, pp. 349–480. MR: 2415378. Zbl: 1181.37049 (cit. on p. 30).

S. Wiggins (2013). *Global bifurcations and chaos: analytical methods*. Vol. 73. Springer Science & Business Media. MR: 0956468. Zbl: 0661.58001 (cit. on pp. xiv, 55, 64).

D. Wilczak (2003). "Chaos in the Kuramoto-Sivashinsky equations—a computer-assisted proof." *Journal of Differential Equations* 194.2, pp. 433–459. MR: 2006220. Zbl: 1050.37017 (cit. on p. 123).

D. Wilczak (2005). "Symmetric heteroclinic connections in the Michelson system: a computer assisted proof." *SIAM J. Appl. Dyn. Syst.* 4.3, pp. 489–514. MR: 2173545. Zbl: 1120.34033 (cit. on p. 123).

R. F. Williams (1979). "The structure of Lorenz attractors." *Inst. Hautes Études Sci. Publ. Math.* 50, pp. 73–99. MR: 556583. Zbl: 0484.58021 (cit. on p. xiii).

J. A. Yorke and K. T. Alligood (1983). "Cascades of period-doubling bifurcations: a prerequisite for horseshoes." *Bulletin of the American Mathematical Society* 9.3, pp. 319–322. MR: 0714994. Zbl: 0541.58039 (cit. on p. 71).

Index

Títulos Publicados — 32º Colóquio Brasileiro de Matemática

Emergence of Chaotic Dynamics from Singularities – *Pablo G. Barrientos, Santiago Ibáñez, Alexandre A. Rodrigues e J. Ángel Rodríguez*

Nonlinear Dispersive Equations on Star Graphs – *Jaime Angulo Pava e Márcio Cavalcante de Melo*

Scale Calculus and M-Polyfolds An Introduction – *Joa Weber*

Real and Complex Gaussian Multiplicative Chaos – *Hubert Lacoin*

Rigidez em Grafos de Proteínas – *Carlile Lavor*

Gauge Theory in Higher Dimensions – *Daniel G. Fadel e Henrique N. Sá Earp*

Elementos da Teoria de Aprendizagem de Máquina Supervisionada – *Vladimir Pestov*

Función Gamma: Propriedades Clásicas e Introducción a su Dinâmica – *Pablo Diaz e Rafael Labarca*

Introdução à Criptografia com Curvas Elípticas – *Sally Andria, Rodrigo Gondim e Rodrigo Salomão*

O Teorema dos Quatro Vértices e sua Recíproca – *Mário Jorge Dias Carneiro e Ronaldo Alves Garcia*

Uma Introdução Matemática a Blockchains – *Augusto Teixeira*